STEP-BY-STEP GUIDE: CARBURETOR TUNEUP & OVERHAUL

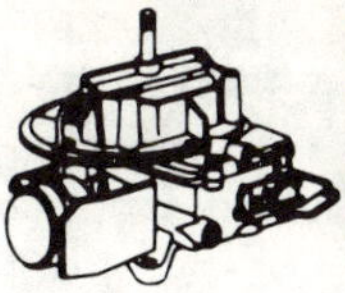

STEP-BY-STEP GUIDE: CARBURETOR TUNEUP & OVERHAUL

BY HERSCHEL & RAY WHITTINGTON

TAB BOOKS Inc.
BLUE RIDGE SUMMIT, PA. 17214

FIRST EDITION

SEVENTH PRINTING

Printed in the United States of America

Hardbound Edition: International Standard Book No. 0-8306-5814-9

Paperbound Edition: International Standard Book No. 0-8306-4814-3

Library of Congress Card Number: 75-20841

Cover illustration courtesy Pontiac Motor Division.

Contents

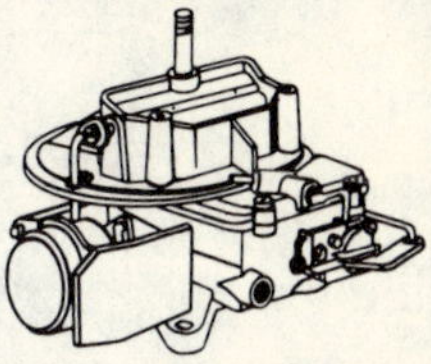

Preface

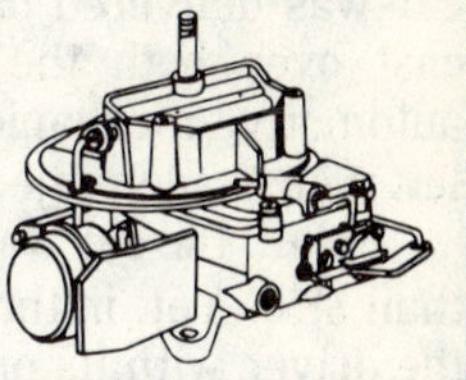

Car maintenance has become a major source of irritation as well as expense for too many people. In a recent consumer report, a Dallas television station recounted an actual experience with what has become an all-too-common possibility—car repair "rip-off!"

They used a two-year-old station wagon—typical of what the average middle-class suburban American housewife might have been driving in 1974. After having its engine overhauled and tuned to a "T", two of the spark plug wires were crossed, which naturally caused the car to run rough.

Then a young lady reporter, wired for sound and followed discreetly by a cameraman in an unmarked van, was dispatched to a variety of automotive service centers.

First, she returned to the agency that had originally sold the car. There she was charged $32.50 for a minor tuneup and was told emphatically that her car must have a valve job—estimated cost: $140. She left this garage with her wires still crossed.

Next, she pulled into a small independent service center. Here, the service manager spotted her problem immediately, corrected it, and sent her on her way without charge.

Then she stopped at a major oil company service center where, after extensive testing, her car received another minor tuneup at a cost of more than $35. This time, however, in replacing the spark plugs, the crossed wires were inadvertently uncrossed.

After recrossing the wires, she drove to another independent one-mechanic garage. This time, a carburetor overhaul was performed and a valve job recommended. The cost: nearly $50; the result: the car still ran as rough as before.

Finally, our lady reporter took the car to another service center. Here, for the fourth time in four days, the car received a minor tuneup. Again she was charged nearly $40 and again the car was delivered back to her. Needless to say, a shadow was cast over both the competence and the integrity of Dallas automotive mechanics, and indeed all mechanics, as this report left their trade subject to serious doubt.

In all, the car was tied up for three full days at a cost of more than $150. Yet, in three out of five tries, the car was returned to the driver with its problem unsolved. Incidentally, in Dallas, all auto repair shops must be licensed by the city; some call this license a "permit to do business," but many recognize it more as a "permit to do larceny." Let us hasten to suggest, however, that we are convinced that this poor showing resulted more from ignorance than dishonesty.

Without a doubt the majority of mechanics are capable and honest. Unfortunately, however, there are too many who, for one reason or another, seem to be wholly incompetent. But, regardless of their qualifications, all stay busy, for there is no end to the work that must be done if our millions of automobiles are to keep rolling. Just a few days ago I heard one harried service manager confirm that the only question he asks a prospective mechanic employee any more is, "Do you have your own tools?"

But, even if all mechanics were both technically qualified and honest, often you would be far ahead if you could do your car maintenance and repairs yourself. Not only could you save yourself time, but money. For instance, last summer my station wagon showed definite signs of needing a brake job—after 50,000 miles the brakes were noisy, and they no longer gave me that "feeling of confidence" we all like to experience when we step on a brake pedal. So I wheeled into one of those franchised brake specialty shops for a "free" estimate. And, although this was before the recent phenomenal increase in brake-part costs, the estimate I got was $123.65. Needless to say, this figure seemed a bit excessive to me, so I decided to do it myself.

Now, of the team that put this book together, the other member is the master mechanic—I'm just the writer. And the two of us live 300 miles apart, so I was on my own, and I'd not done a brake job in nearly a quarter of a century. Nevertheless, I went to a neighborhood parts store where I purchased the parts, tools, and materials I had to have for this task. Of course, I had to take my brake drums to an automotive machine shop to have them turned, but everything else I did myself.

Not counting the three-hour wait for my shoes, the entire task consumed a mere hour and a half. But, more importantly, it cost me only a few pennies more than $27 for a saving of nearly 80%.

You don't have to be a mechanic, or even be "mechanically inclined," to save this kind of money on repair bills. All you have to be capable of is reading and following the easy, simplified step-by-step instructions contained here.

Incidentally, please treat this book as a part of your working materials—break the spine if necessary, so that it will stay open at the right place while you're working, or even tear out the pages if that will make your work easier.

Our special thanks to Marvel-Schebler, Division of Borg-Warner Corp., for allowing us to reproduce instructional material from their carburetor overhaul kits.

Herschel Whittington.

Chapter 1

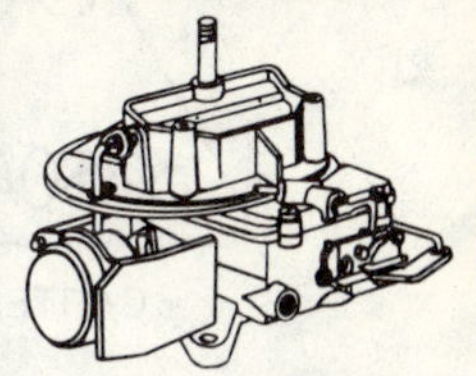

Tools and Materials

The tools you will need to tune up any and all of the carburetors discussed in this book are illustrated in Fig. 1-1 and described below:

- Socket assortment.
- Long-nosed pliers.
- Adjustable wrench with at least a 1 in. span.
- Open-end or box-end wrenches, 9/16 in. × 3/8 in. and 7/16 in. × 3/8 in.
- Phillips screwdriver, medium bit.
- Flat screwdrivers, 7/16 in. bit and 3/16 in. bit.

In addition, you will need enough carburetor cleaner to clean all parts simultaneously. Use only properly labeled cleaner and heed all directions, cautions, and warnings. Recognized brands include Bendix, Gunk Hydro-Seal, and Delco X-55. If the cleaner is supplied without a parts tray, obtain a *metal* bucket to hold the parts. The bucket should be spotlessly clean.

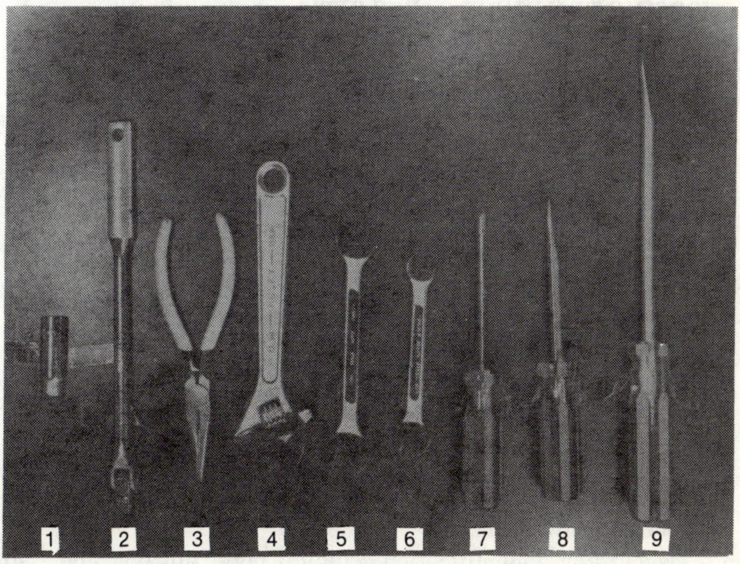

Fig. 1-1. Carburetor work requires a minimum investment in tools.

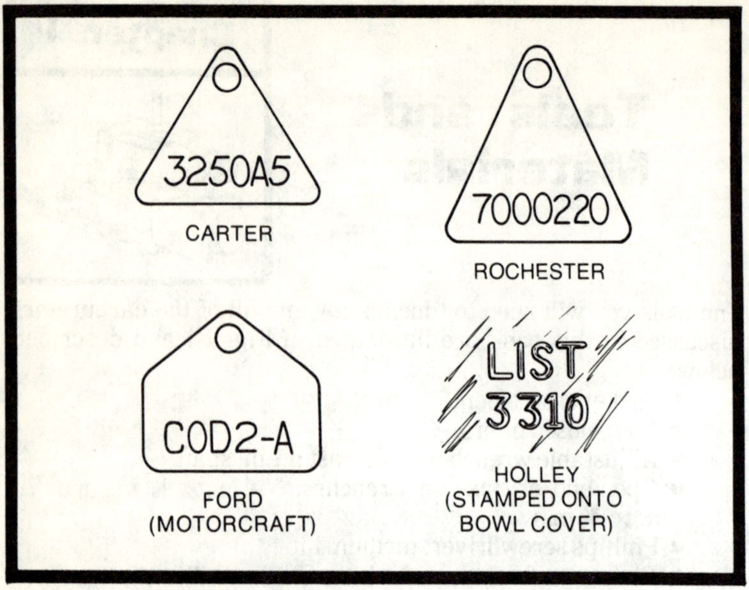

Fig. 1-2. Typical carburetor identifications. Many late-production carburetors employ cast or stamped part numbers rather than detachable tags.

A half hour's soak is usually enough, although severely corroded carburetors may require longer. Flush the parts with water or kerosene to remove all traces of suspended dirt.

CHOOSING THE RIGHT PARTS KIT

Many good tuneup kits are available wherever car parts are sold. To be sure you get the right kit for your carburetor, have the following information when you go to the parts store:

- Year, make, and model of your car.
- Engine size (number of cylinders, horsepower rating, and cubic inches of displacement if known).
- Brand name and identification number of your carburetor. (Carburetors have cast-on or attached nomenclatures. See Fig. 1-2).
- Number of barrels in your carburetor. (Remove the air cleaner and count the barrels.)

Note: Each carburetor kit accommodates several different models; you may have parts left over after assembly. Don't worry about it.

Note: Always use new gaskets.

Further Identification Hints

Carburetor identification may not always be so straightforward. Some mechanics hang the ID tags on a nail to keep count of how many carburetors they overhaul, and kids regularly replace Quadrajets and 4GCs with Holleys.

The list below is necessarily incomplete, but it is a starting point to help you and your counterman find the right kit. In the last analysis you may have to consult the car maker's manual for the correct carburetor make and number. Carburetor manufacturers provide interchange lists which are useful if a nonfactory substitution has been made.

Carter One-Barrel Carburetor, Model BBS.

6-cyl Dodge	1961-1971
6-cyl Plymouth	1954-1971
6-cyl Dodge (truck)	1961-

Carter One-Barrel Carburetor, Model YF.

6-cyl Chevrolet	1963-1967
6-cyl Ford (pass. car & truck)	1968-
6-cyl GMC (truck)	1963-1967

Carter Two-Barrel Carburetor, Model BBD.

8-cyl Chrysler	1955-1967
8-cyl Dodge	1959-
8-cyl Plymouth	1959-

Carter Quadrajet Four-Barrel Carburetor: Replacement only.

Carter Four-Barrel Carburetor, Model AFB.

8-cyl Buick	1957-1967
8-cyl Chevrolet	1958-1965
8-cyl Chrysler	1959-1967
8-cyl Dodge	1958-1971
8-cyl Ford	1957-1960
8-cyl Plymouth	1958-1971
8-cyl Pontiac	1957-1967

Ford One-Barrel Carburetor, Model F-1.

6-cyl Ford (pass. car & truck)	1963-1969
6-cyl Mercury	1963-1969

Ford Two-Barrel Carburetor, Model 2100.

8-cyl Ford (pass. car & truck)	1963-
8-cyl Mercury	1963-

Ford Four-Barrel Carburetor, Model 4100.

8-cyl Ford	1957-1969
8-cyl Mercury	1958-1969

Ford Four-Barrel Carburetor, Model 4300.

8-cyl Ford	1966-
8-cyl Lincoln	1968-
8-cyl Mercury	1967-

Holley One-Barrel Carburetor, Model 1904-1920.

Some American Motors cars	
6-cyl Dodge	1960-
6-cyl Ford (pass. car & truck)	1952-1967
6-cyl Plymouth	1960-

Holley One-Barrel Carburetor, Model 1909. Replacement only.

Holley Four-Barrel Carburetor, Model 3310.

8-cyl American Motors (some)	1958-1970
8-cyl Chevrolet	1964-1972
8-cyl Dodge	1960-1972
8-cyl Ford	1958-1971
8-cyl Ford (truck)	1957-
8-cyl Mercury	1958-1971
8-cyl Plymouth	1960-1972

Rochester One-Barrel Carburetor, Models B, BC & BV.

6-cyl Buick Special	1964-1965
4-cyl Chevrolet	1962-
6-cyl Chevrolet (pass. car & truck)	1950-1967
6-cyl Oldsmobile F-85	1964-1967
6-cyl Pontiac (Canadian)	1955-1967
4-cyl Pontiac Tempest	1961-1963
6-cyl Studebaker	1965-1966

Rochester Monojet One-Barrel Carburetor, Model MV.

6-cyl Buick Special	1968-
6-cyl Chevrolet (pass. car & truck)	1968-
6-cyl GMC (truck)	1968-
6-cyl Oldsmobile F-85	1968-
6-cyl Pontiac Firebird	1968-
6-cyl Pontiac Tempest	1968-1972

Rochester Two-Barrel Carburetor, Models 2G, 2GC & 2GV.

8-cyl Buick	1961-
8-cyl Chevrolet (pass. car & truck)	1955-
8-cyl Oldsmobile	1955-
8-cyl Pontiac	1955-
8-cyl Studebaker	1965-

Rochester Quadrajet Four-Barrel Carburetor, Model 4MV.

8-cyl Buick	1966-
8-cyl Cadillac	1967-
8-cyl Chevrolet (pass. car)	1967-
8-cyl Chevrolet (truck)	1968-
8-cyl GMC (truck)	1968-
8-cyl Oldsmobile	1966-
8-cyl Pontiac	1966-

Rochester Four-Barrel Carburetor, Model 4GC.

8-cyl Buick	1955-1966
8-cyl Cadillac	1955-1966

8-cyl Chevrolet (pass. car)	1958-1966
8-cyl Chevrolet (truck)	1957-1967
8-cyl Oldsmobile	1952-1965
8-cyl Pontiac	1954-1965

Stromberg Two-Barrel Carburetor, Model WW.

8-cyl Dodge	1953-1967
8-cyl Dodge (truck)	1955-1969
8-cyl Plymouth	1960-1967

Stromberg Two-Barrel Carburetor, Model WWC.

8-cyl Chrysler	1961-1967
8-cyl Dodge	1960-1967
8-cyl Plymouth	1965-1967

The listing is meant to be nothing more than a guideline. And it can serve that purpose only in a limited way. For example, if you drive a 1966 8-cylinder Plymouth you may find your car listed under several carburetor types—that's because 1966 8-cylinder Plymouths used several different carburetors, usually for different engine sizes or performance requirements. On the other hand, if your car is a 1973 4-cylinder Chevrolet, you won't find it listed anywhere. Why? Because it probably has a carburetor of foreign design and manufacture. So...we must leave it to you, in the end, to discover for yourself the nomenclature of your carburetor.

CAUTIONS

We would prefer not to have to use the word "caution" anywhere in this book, for it might frighten off some of you, and our purpose is to encourage you—not discourage you. We want you to tackle this fairly simple and quite safe procedure. But, like any human endeavor, it does carry with it some potential hazards for those who either do not know or ignore the fact that gasoline is volatile, or that carburetor cleaning solvent can be harmful to human tissue, or that metal carburetor parts propelled by compressed air can be as dangerous as projectiles fired from a pellet gun. So, we offer these three cautions:

Do not use gasoline for any purpose except that for which it is sold: to operate your car's engine. And never have an open flame around gasoline—not even a cigarette.

Do not leave carburetor cleaning solvent on your skin for any length of time; wash it off with water. And if you get some solvent in your eye, wash it out immediately.

Do not use compressed air to dry carburetor parts or blow out vents and passages. The reason is simple enough: if you overlook one of the check balls or needle valves, it can become a dangerous projectile.

Chapter 2

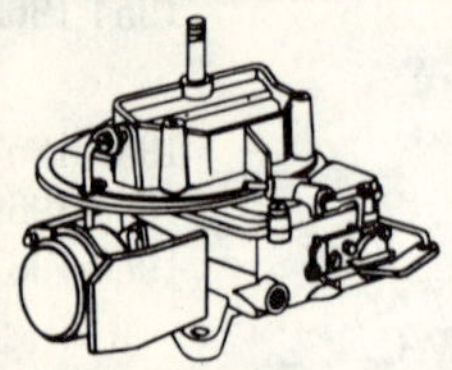

Carburetor and Fuel System Basics

The carburetor mixes gasoline with air to form an explosive vapor. While we tend to think of the carburetor in terms of the fuel which passes through it, the amount of air "inhaled" is fully as critical as the fuel. Best engine power is produced with air/fuel ratios on the order of 15:1 by weight. Each gallon of gasoline requires some 1100 cubic feet of air for vaporization. Mixing large amounts of air with very small amounts of gasoline requires great precision and is responsible for the mechanical complexity of modern carburetors. The Quadrajet, for example, is an assembly of 208 parts.

While this book describes overhaul and tuneup procedures for most popular domestic carburetors (accounting for some 95% of the carburetors in use today), we cannot describe them all. Complete coverage would involve approximately 4000 makes and models, an impossible task. However, while the hardware varies, the basic principles are the same for all. The Weber DCOE, perhaps the most sophisticated carburetor ever built for land vehicles, and the humble Rochester single-barrel, all function by the same physical laws. If you understand one, you understand them all.

CARBURETOR FUNCTIONS

The carburetor has three major functions. The first is to prepare the fuel for vaporization within the engine. Air and fuel are mixed in two stages. Initially raw gasoline is converted into an emulsion or froth. It remains liquid, but broken by tiny air bubbles. The next reduction is atomization. The fuel is converted to a fine spray as it leaves the jets. Finally the fuel is vaporized inside of the engine by heat from the combustion chamber.

The carburetor's second big job is to meter the air and fuel to match engine requirements. During cold starts and at idle the engine needs a gasoline-rich mixture. At cruise the mixture should be leaned with air for economy. And at full power the engine again demands a rich mixture. All of this is done automatically.

The third function is to throttle the engine by controlling the amount of air/fuel mixture leaving the carburetor.

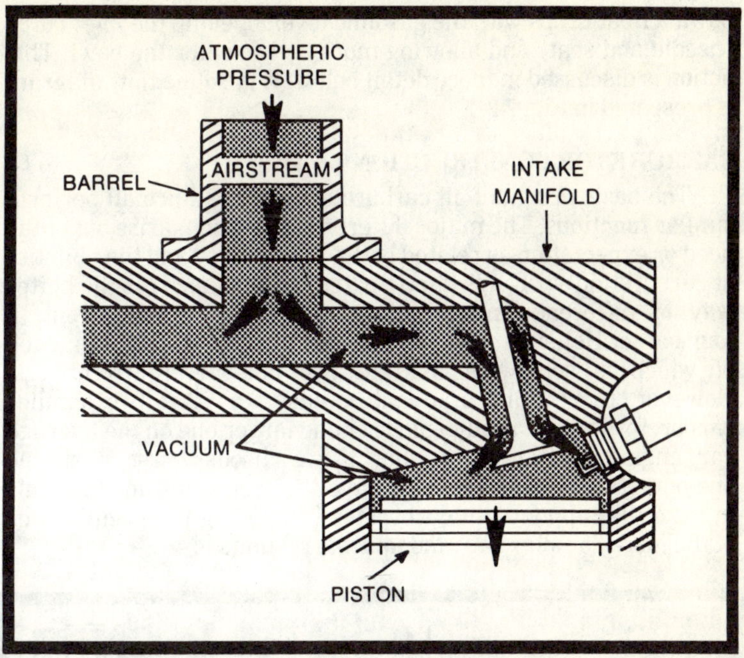

Fig. 2-1. Air flow on the intake stroke.

Air Entry

Except in those rare situations where superchargers are used to push air into carburetors, it is more appropriate to say that air is "sucked" into a carburetor. When an engine piston moves downward in its cylinder on the *intake stroke*, it creates a partial vacuum behind it. Because "nature abhors a vacuum" and because there is no other way for air to enter this partially vacated cylinder, air rushes in through the carburetor throat via the intake manifold and through the open intake valve to the cylinder. The amount of air inducted is a function of engine displacement and rpm.

For instance, an engine rated at 350 CID (cubic inches of displacement) turning 3000 rpm theoretically should "inhale" nearly 300 cubic feet of air mixed with gasoline each minute. However, an engine is rarely this efficient, and the actual consumption would be more on the order of 100 cubic feet of air and atomized fuel per minute.

Gasoline Entry

Unlike air, which is sucked into the carburetor, gasoline arrives there under pressure. A *fuel pump*, situated in the fuel line between the tank and the carburetor, lifts gasoline from the tank and delivers it to the carburetor under pressure. Fuel pumps operate at between 4 and 6 psi. As gasoline is used from the fuel bowl, a float drops with the gasoline level, opening the *inlet valve* (needle and seat) and allowing more fuel to enter the bowl. This action is discussed in more detail below. A gasoline flow diagram is presented in Fig. 2-2.

CARBURETOR CONSTRUCTION

The basic design of all carburetors is alike since all perform similar functions. The major differences in design arise out of the need or expectation as related both to the quality and the quantity of carburetion—perhaps "anticipated performance" is the better way of differentiating these design needs. In any event, a four-barrel Holley is much like two two-barrel Rochesters, each of which is somewhat akin to two one-barrel Strombergs. However, there is no practical or easy way that two smaller carburetors can be used in place of one larger one on the average car engine. Our point is that if we discuss first a simple one-barrel carburetor, we will have the technical fundamentals in hand that apply to all carburetors, including the peculiarities of the more complex two- and four-barrel units.

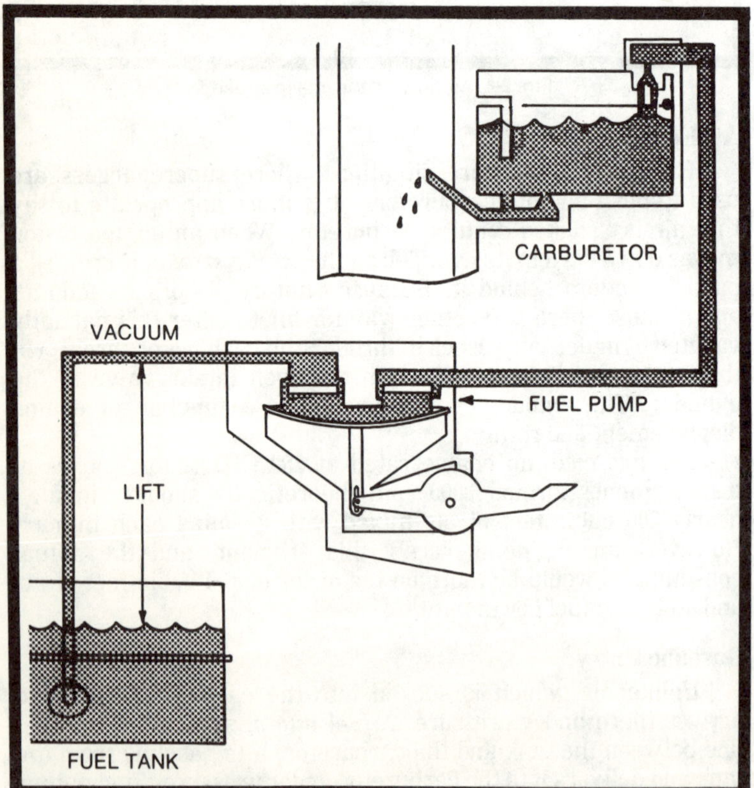

Fig. 2-2. Fuel flow. The pump generates negative and positive pressures.

First of all, a carburetor must mount to an engine. For this reason, every carburetor must have a *base*. Usually this base is a separate casting, although it may, on some smaller units, be integral with the carburetor *bowl*. Figure 2-3 illustrates typical carburetor castings.

The second major casting is called the carburetor *bowl*. Not only does it house a bowl for constant-level storage of gasoline, it houses most of the bits and pieces that work together to make up the various carburetor systems. All the intricate metering, emulsification, and atomization occur here.

The third common casting is the bowl cover. Its principal task is to cover the bowl while providing access to the hardware contained in the bowl. Some of this hardware may also be in the cover.

Carburetor castings and parts are made of zinc-based alloy, aluminum, cast iron, and plastic, depending upon their functions. Generally speaking, all are tough and resistant to the corrosive

Fig. 2-3. Basic carburetor castings.

effects of gasoline. Even so, a carburetor is a busy place, and all those tons of air sucked through bring with them many pollutants (such as dust) that gradually cause wear, both to the carburetor and to the engine. But the most troublesome pollutants are those in the gasoline and those in the air above the gasoline in the fuel bowl. Some of these pollutants appear as deposits on metering parts, such as the needle and seat. Others attack the gaskets and other nonmetallic parts, shrinking, warping, cracking, and dissolving them. Thus, sooner or later, all carburetors must be overhauled, though one of the miracles of modern technology is that such a busy unit can perform as well as it does for as long as it does.

CARBURETOR SYSTEMS

A carburetor is a collection of five major systems:

1) *The float system* maintains a proper level of gasoline in the carburetor fuel bowl.
2) *The main metering and power system* channels gasoline from the fuel bowl to the carburetor throat for normal road driving.
3) *The accelerator pump system* provides a flood of fuel to the carburetor throat when the accelerator pedal is first depressed.
4) *The idle system* maintains a smooth, steady engine speed at idle, even when the engine is loaded by accessories such as the air conditioner and alternator.
5) *The choke system* restricts the flow of air into the carburetor throat to accommodate cold starts and cold driveaway.

We will discuss each system in turn.

The Float System

As we've noted, gasoline enters the carburetor through a valve called the needle and seat. This valve is controlled by a float, which in turn is controlled by the level of gasoline in the fuel bowl. Thus, we've identified the three major items in the float system:

- Fuel bowl
- Float
- Needle and seat

These parts are shown in Fig. 2-4.

Whenever gasoline is drawn from the fuel bowl, the float drops, allowing the pressure created in the gasoline line by the fuel pump to force the needle away from the seat and permitting gasoline to flow through the seat into the fuel bowl. As soon as the fuel level again raises the float sufficiently, the needle is forced

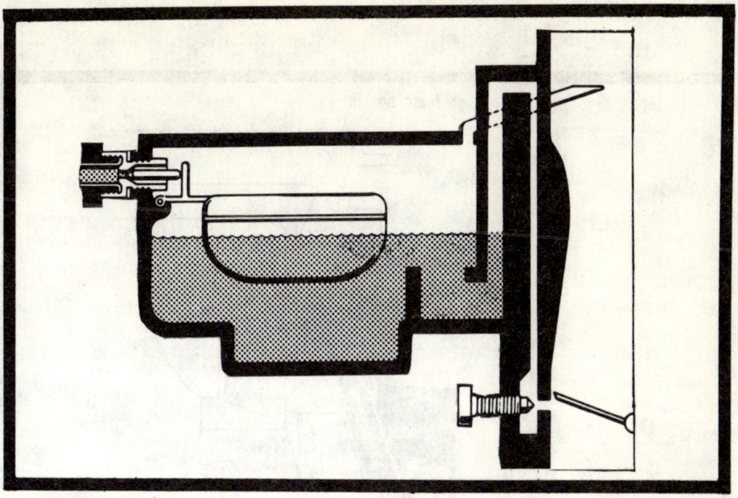

Fig. 2-4. Fuel bowl, float, and needle and seat. The float level fixes the carburetor's internal fuel level.

back into the seat by the float arm, and gasoline flow into the bowl ceases. At steady highway speeds, flow out of as well as into the fuel bowl is continuous, and the level remains constant. Other factors such as engine vibration, angle of climb, and side tilt on turns affect fuel levels on carburetors not built to combat such influences.

Of course, at no time is the fuel bowl filled only with gasoline—there is always a pocket of air above the gasoline. The pocket of air becomes larger and smaller as the gasoline level falls and rises. A vent is provided in the bowl cover to permit a free flow of air in and out of the fuel bowl. Since air contains some moisture as well as other foreign materials and because this air is not always filtered, the vent may become a source of pollutants. But it is a necessity whose value far exceeds its shortcomings. And because, in a sense, gasoline boils all of the time and especially in a hot engine compartment, the vent allows the pressures created by such boiling to be dissipated. Otherwise vapor pressure would become so great that gasoline would gush out of the bowl into the carburetor throat without being properly emulsified and atomized, flooding the engine.

Air rushing through the throat of the carburetor tends to lose pressure at the point of fuel entry; to prevent a *major* difference in fuel-bowl and throat pressure, a second vent—called the *balance vent*—opens to the lip of the carburetor throat on some designs. Air passing into the fuel bowl through this vent during normal engine operation will have passed through the air filter.

Little more needs to be said about the float system except to mention the float itself. Needless to say, its proper level setting is vital to operation of the carburetor and the car itself. Float

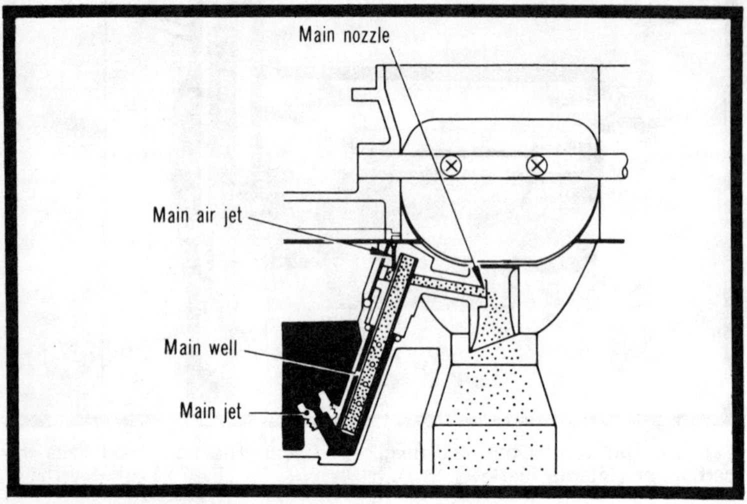

Fig. 2-5. Fixed-diameter main jet. This jet provides fuel for cruise and is calibrated somewhat leaner than the others. For fuel-economy considerations the main jet emulsifies the fuel before discharge into the carburetor bore. (Courtesy Chrysler Motors Corp.)

settings can go astray due to wear of the pivotal pin (or hinge), because of gasoline seepage into the older hollow metal types, or saturation of the plastic types used in many late-model cars. Also, floats may function inadequately if the float-assist spring has lost tension.

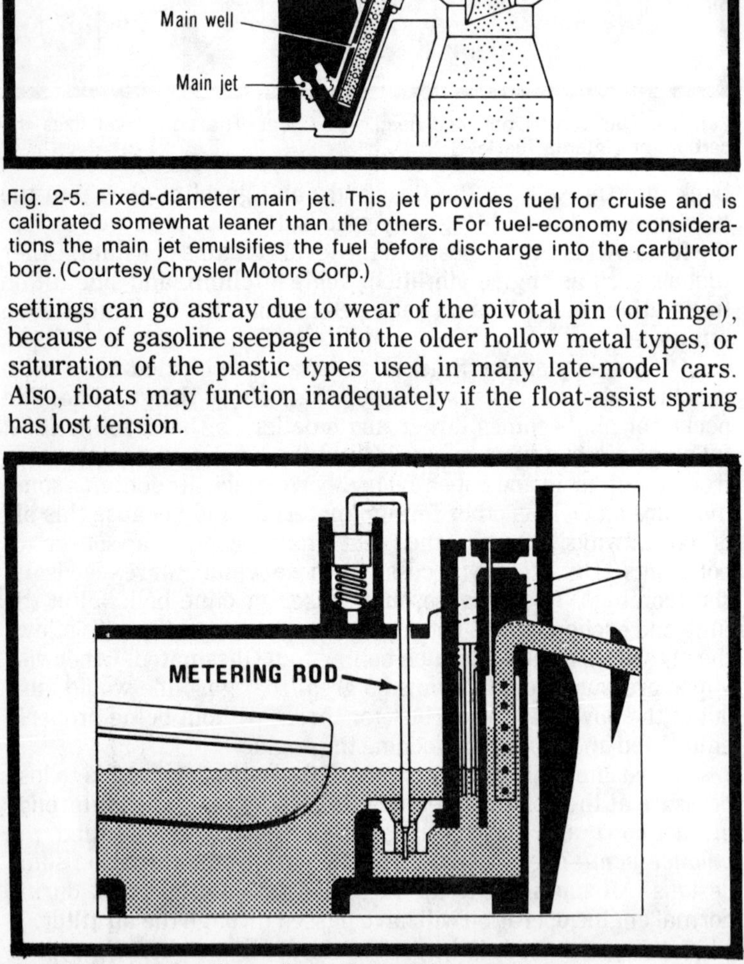

Fig. 2-6. A variable-orifice main jet. The rod is retracted out of the jet for more fuel at high speed or under load. The metering rod is retracted by means of a vacuum-sensitive piston.

The Main Metering and Power System

The *main-metering jet* provides fuel for normal road operation. Typically it kicks in at about 25 mph. The jet may have a fixed orifice (Fig. 2-5) or may be variable (Fig. 2-6). In the latter case the jet size is controlled by a tapered or stepped *metering rod*. At low speeds and under light loads the large diameter of the rod is in the orifice. As demand on the engine increases the rod retracts to allow more fuel to flow through the jet. Metering rods may be operated by vacuum (as explained below in the discussion of the power jet) or they may be tied mechanically to the throttle valve.

Regardless of the jet type, fuel passes through it into the fuel *well* where it comes into contact with air and is emulsified prior to discharge. Various systems are used to introduce a controlled amount of air bleed, as you can see from the illustrations.

Fuel is discharged in the carburetor throat at its smallest diameter—the *venturi* (Fig. 2-7). Some two hundred years ago Italian physicist G. B. Venturi discovered that a fluid flows faster through a narrowed-down section of a tube than through its larger diameters. This increase in velocity is accompanied by a decrease in pressure.

Fuel in the carburetor bowl is at atmospheric pressure because of the vent. The pressure differential between the fuel bowl and the venturi causes flow through the main metering jet, discharge nozzle, and into the carburetor throat.

In addition to the primary venturi formed by narrowing the bore, one or more booster venturis may be present. These

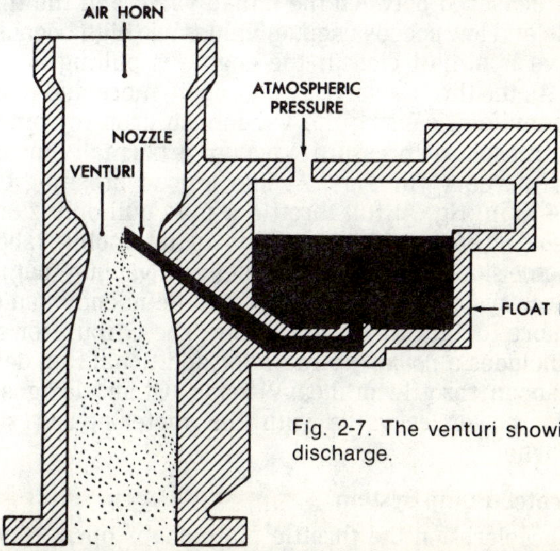

Fig. 2-7. The venturi showing fuel discharge.

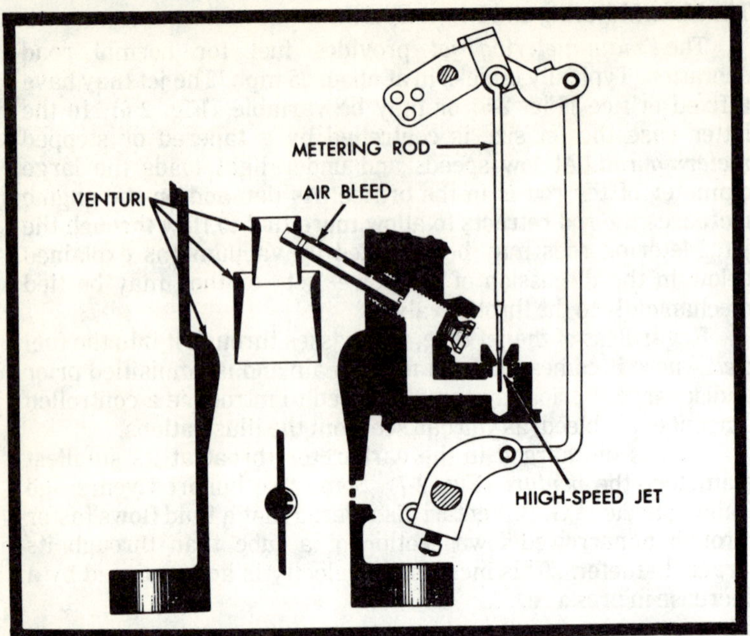

Fig. 2-8. Triple venturi cluster in combination with a mechanically operated metering rod.

boosters are concentric with the main bore (Fig. 2-8). The purpose is to make the main metering system operation more efficient without reducing air flow at wide-open throttle. Manifold vacuum is the key to its operation. This is negative pressure or vacuum as measured between the intake valve and the throttle plate. At idle and low speeds vacuum values are high because the throttle valve is almost closed; the engine is pulling against a restriction. As the throttle valve swings open, more and more air enters the manifold, causing the vacuum to drop to something just below atmospheric pressure. Vacuum is normally measured in inches of mercury (in. Hg). At idle a good presmog engine will pull 22–23 in. Hg. At full throttle it may pull only 2 or 3 in. Vacuum also decreases with load. When an automobile labors up a hill the engine slows and loses efficiency as a vacuum pump.

The power system senses this vacuum deficiency and opens to deliver more fuel for increased power. The carburetor shown in Fig. 2-9 includes a *power jet valve* for additional fuel delivery under wide-open throttle or heavy loads. Other designs may combine the power jet valve with the metering-rod system discussed earlier.

The Accelerator-Pump System

Under acceleration the throttle plate snaps open. Since air weighs less than gasoline it has less inertia. Consequently, the air

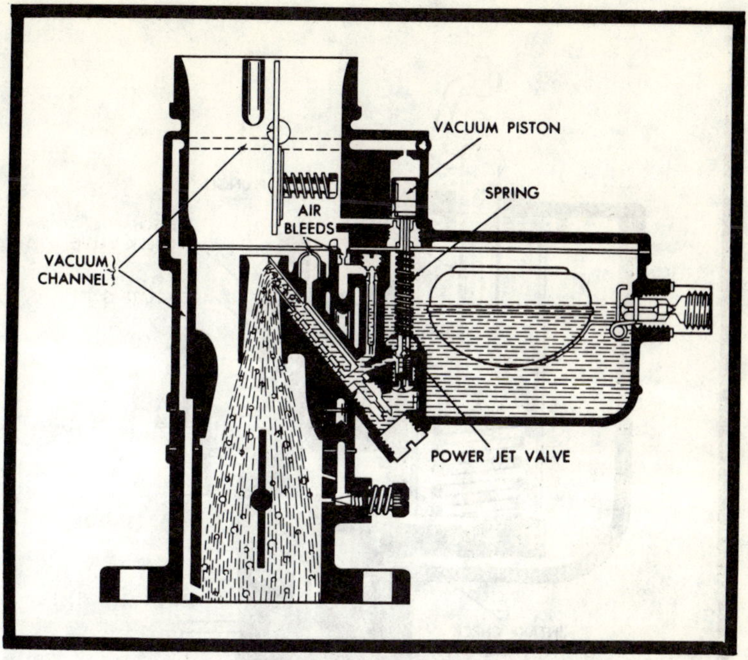

Fig. 2-9. Power jet valve.

stream responds more quickly than the fuel. If uncorrected, the carburetor would go momentarily lean until the fuel caught up with the air stream. The *accelerator pump* compensates by delivering a shot of raw gasoline into the carburetor bore. The point of entry is above the venturi.

Fuel enters the pump chamber through a channel at the base of the float bowl (Fig. 2-10). It passes through a check valve and is stored in the pump chamber. When the throttle is depressed, the pump element—a plunger or diaphragm—displaces fuel from the pump chamber through a second check valve to the discharge(s) in the carburetor bore. The inlet check valve prevents fuel from backing up into the float bowl; the outlet valve prevents "pullover" of fuel out of the pump chamber due to carburetor vacuum. Leaking outlet check valves are responsible for many cases of poor fuel economy.

The Idle System

At low speeds air velocity through the venturi is low and little vacuum is developed. The main metering and power system is dormant, since it is entirely a creature of venturi vacuum. In addition, gasoline engines are reluctant to idle and must be bribed with a richer-than-normal mixture. These two considerations have caused engineers to add a second fuel-delivery system to the carburetor, independent of the first.

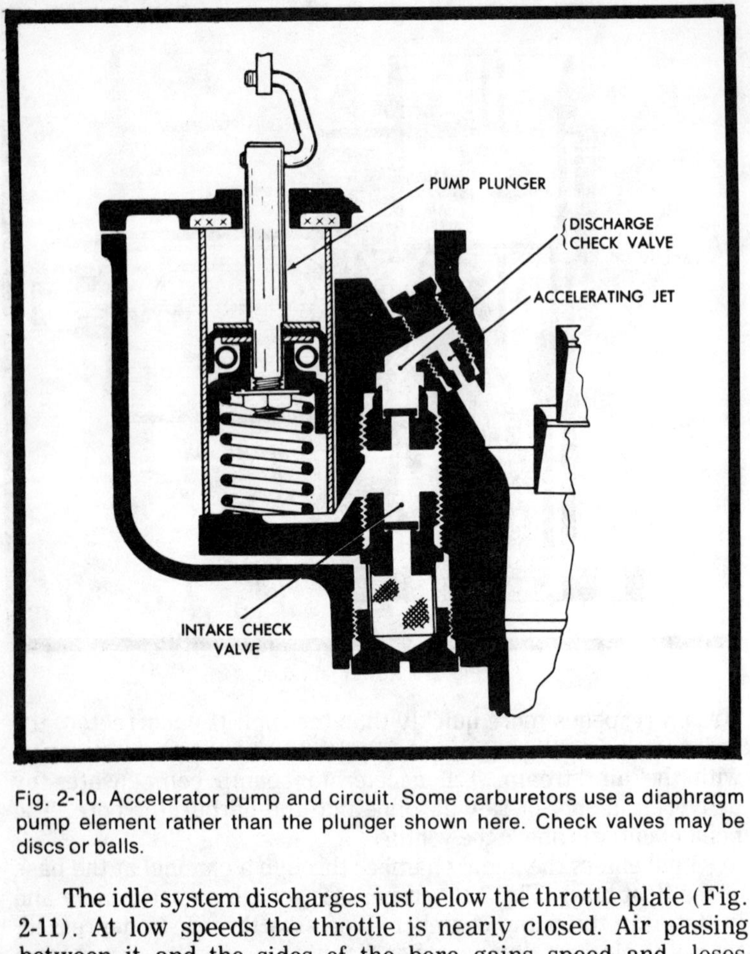

Fig. 2-10. Accelerator pump and circuit. Some carburetors use a diaphragm pump element rather than the plunger shown here. Check valves may be discs or balls.

The idle system discharges just below the throttle plate (Fig. 2-11). At low speeds the throttle is nearly closed. Air passing between it and the sides of the bore gains speed and loses pressure. The throttle plate is an obstruction in the bore and is therefore a kind of venturi. In response fuel leaves the discharge port. A second and sometimes a third port is positioned above the first. These are known as the off-idle or low-speed ports. They make the transition between idle and part-throttle operation smoother. The idle mixture is adjustable from outside of the carburetor by means of a screw.

The Choke System

American carburetors employ a *choke valve* mounted above the venturi. If it were not for its location, one could mistake it for a throttle valve. During cold starts the choke swings closed, either on command from the driver or automatically. Very little air gets into the bore. Consequently pressure drops below the choke. The idle and main metering systems flow in response to

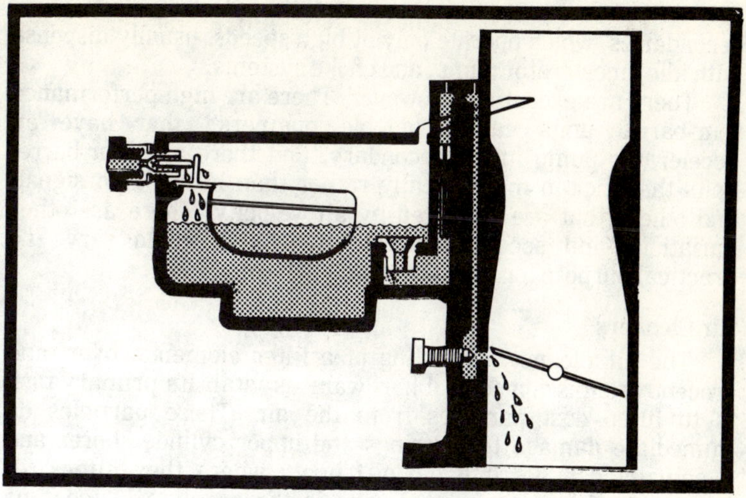

Fig. 2-11. Basic idle system.

this partial vacuum, causing a very rich mixture to be inducted into the engine (Fig. 2-12).

Automatic choke mechanisms will be discussed in a following section.

Multibarrel Carburetor

Carburetors are normally built with one, two, or four throats or barrels. Two-barrel carburetors usually have one fuel bowl with a dual system that feeds the two barrels simultaneously. Four-barrel carburetors function as two-barrel carburetors until manifold pressure triggers the secondary barrels. The

Fig. 2-12. When the choke is closed idle and main-metering systems flow.

secondaries, which operate only at high speeds, usually dispense with idle, accelerator pump, and choke systems.

There are exceptions, however. There are high-performance four-barrel units called "double pumpers" that have an accelerator pump in the secondary, and there are four-barrel units that kick-in mechanically rather than by vacuum signal, and others that are triggered by air velocity. There are other variations and special applications, but it would serve no practical purpose for us to discuss them here.

Air Cleaners

The air cleaner, consisting of a filter element, cover, and (recently) emission-control hardware, is vital. Its primary task is to filter dust particles from the air. These particles do immediate damage to the rings and upper cylinder bores and eventually can get into the oil supply where they attack all bearing surfaces. Once an engine is thoroughly "dusted," no amount of cleaning or flushing can remove the particles. They remain imbedded in the bearing metal.

The condition of the filter affects carburetion. A clogged filter will increase fuel consumption since it increases vacuum in the carburetor. The engine runs as if it were partially choked. Removing the element will not make the engine run easier, at least on late-model cars and trucks. Nor do modifications to the cover, e.g., cutting additional air holes or sawing off the snorkel, help breathing.

Three filter elements are in common use. The oldest and least effective of the three is the *oil-bath* type. It is identified by the oil reservoir below the wire-mesh element. Incoming air must change direction as it enters and leaves the element causing dust particles to be centrifuged out where they are trapped in the oil sump. These filters should be cleaned in solvent and reoiled periodically. *Polyurethane* elements are made of a sponge-like plastic. They should be cleaned and wetted with oil per the manufacturer's instructions. Too much or too heavy a grade of oil will cause the carburetor to go rich. *Pleated-paper* elements are generally conceded to be the best choice for normal road use. Once "broken-in" with a light coat of dust, these elements are virtually 100% efficient. The disadvantage is that they cannot be effectively cleaned. Reverse blowing with compressed air removes caked dust, but particles remain imbedded in the fibers. Solvents or water ruin the element on contact. Inspect the seals (they must be soft and resilient), the pleats for cracks, and the area around the PCV tube for oil stains. If more than 25% of the filter is wetted, discard it. Otherwise change the filter at the suggested mileage interval.

Most late-model automobiles have *thermostatically controlled air cleaners* (Fig. 2-13). The purpose is to stabilize

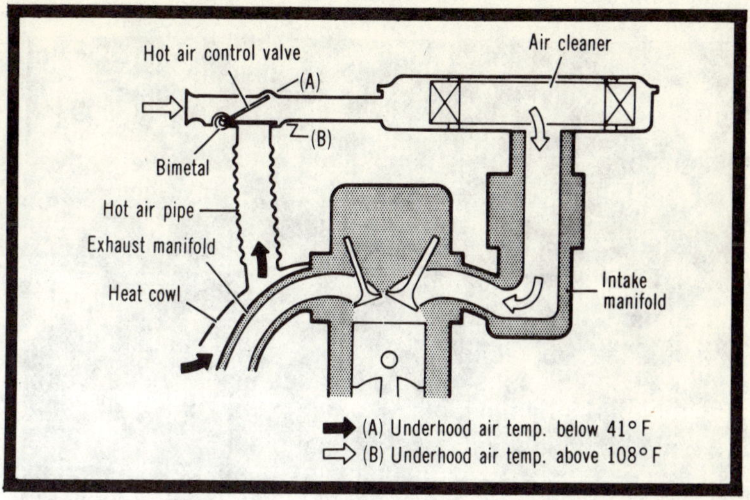

Fig. 2-13. Thermostatically controlled air cleaner. (Courtesy Chrysler Motors Corp.)

carburetion for precise mixture control and to prevent icing. In normal operation, all carburetor air comes into the cleaners from the engine compartment through the snorkel. However, when the engine is cold a thermostat inside the snorkel closes a flap valve in the snorkel throat. This valve shuts off engine-room air, but allows air to be drawn in through a connecting hose from the neighborhood of the exhaust manifold. As temperatures rise the flap valve opens until all air to the carburetor comes from the engine compartment.

To check, take a look at the flap valve in the snorkel before engine is started; it should be closed unless air temperature is up around 100°F. Then start the engine and watch the valve; it should gradually swing open. If it does not open and close properly, check the valve hinges and the vacuum hose. If these are in good condition, you may need to replace the thermostat sensor or the vacuum motor in the snorkel.

Carburetor Solenoids

Another emission control device on late-model cars is the carburetor solenoid (Fig. 2-14). Its purpose is to keep the engine from dieseling—running after the ignition is turned off. Dieseling occurs because a hot cylinder fires by its own heat, which, in turn, pulls air into the next cylinder, and so on, as the engine continues to run and clatter. To prevent this, the solenoid allows the throttle to close completely when the ignition circuit is dead. No air can pass through the carburetor and the engine must stop.

If your car continues to diesel after all else is in tune, the remedy should be simple. Check the solenoid linkage for freedom

Fig. 2-14. Antidieseling solenoid.

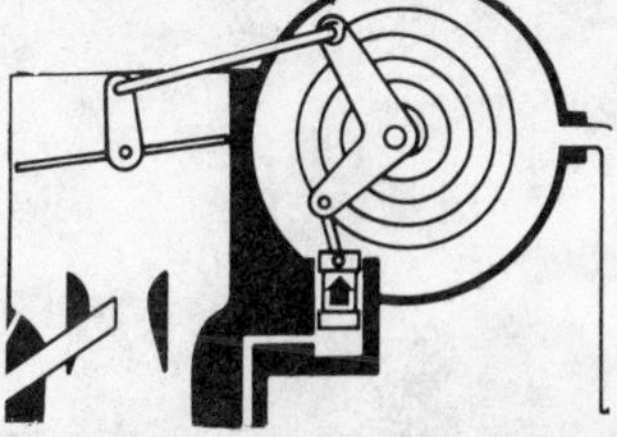

Fig. 2-15. Automatic choke. As the spring is heated it releases tension on the choke butterfly, allowing the vacuum piston to pull the butterfly open.

of movement; check the plunger—it should move forward when engine starts, then fully retract when engine is shut down. Disconnect the wire at the solenoid end, turn on the ignition, and scratch the wire on a metal surface—it should spark. Check the solenoid adjustment at the bracket. The plunger should be able to withdraw far enough for the throttle to close completely.

Automatic Chokes

The automatic choke has one purpose: to provide a rich air/fuel mixture for cold starts and warmups (Fig. 2-15). When the engine is cold the choke automatically closes the butterfly valve(s) in the barrel(s) of the carburetor.

To check choke operation:

When the engine is cold, remove the air cleaner and look at the butterfly; it should be open. Now, depress the accelerator sharply to the floor and release it. This motion should have closed or nearly closed the butterfly.

Start the engine. The butterfly should gradually open as the engine warms up.

If choke does not operate properly:

Check the linkages and pivots for free movement. An aerosol can of choke cleaner may save the day.

If choke opens too slowly or too quickly:

Flooding and excessive fuel consumption occur when choke opens too slowly. When it opens too quickly, the engine may die or run rough. In either case, experiment with the choke adjustment. Loosen the screws and *turn the black housing to the left (counterclockwise) to cause choke valve to open more slowly* (Fig. 2-16). The letters "L" and "R" on the housing stand for "Lean" and "Rich." If the choke housing has lines on it, the single line on the black housing should be lined up with the center line on the metal housing for summer driving. Because of warpage and wear, it may need to be aligned differently. Trial and error is your best approach.

If, after you have tried everything suggested above, your automatic choke still does not work properly, replace it.

Fig. 2-16. Turning the cover tensions or relaxes the spring.

Fuel Pumps

The fuel pump is not, strictly speaking, a carburetor accessory, but it is certainly a vital link in its fuel system (Fig. 2-17). Any malfunction in the fuel pump will upset the carburetor's performance.

Fuel pumps work very hard. Sooner or later they all wear out. And when they do, they can no longer pump gasoline to the carburetors. This trouble is easy to spot: if no gasoline flows into the carburetor, the fuel pump has failed (unless you are out of gasoline, or the fuel line is leaking or plugged). You may need to disconnect the fuel line at the carburetor or at the carburetor side of the fuel pump to be sure nothing comes through when you engage the starter. The best solution for a worn-out pump is to replace it with a new one or a *good* rebuilt one.

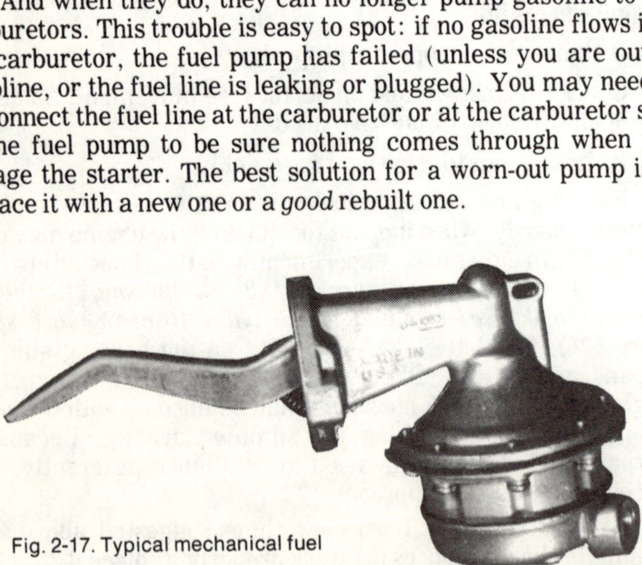

Fig. 2-17. Typical mechanical fuel pump.

Sometimes a fuel pump will deliver at too much pressure, causing the carburetor to flood. The most practical repair is to replace the pump.

The fuel pump should be mounted snugly against the engine, using a new gasket. Fuel-line fittings should be tight. If a fitting leaks gasoline even though it is snug, replace the fitting. When joining to pot-metal castings, turn the fitting at least two threads by hand to prevent cross-threading.

Fuel Filters

The fuel system includes three or more filters at the tank, pump, and between the pump and carburetor. The latter may be spliced into the line or threaded into a boss at the carburetor (Rochester). A clogged filter can cause all sorts of problems—rough running, hard starting, poor fuel economy—which may at first appear to be ignition malfunctions. Replace the pump-to-carburetor filter during each tuneup and be alert to the possibility that the others may have clogged.

Positive Crankcase Ventilation

Since 1963 cars sold in this country have incorporated a positive crankcase ventilation (PCV) system to reduce smog (Fig. 2-18). This was the first example of emission-control

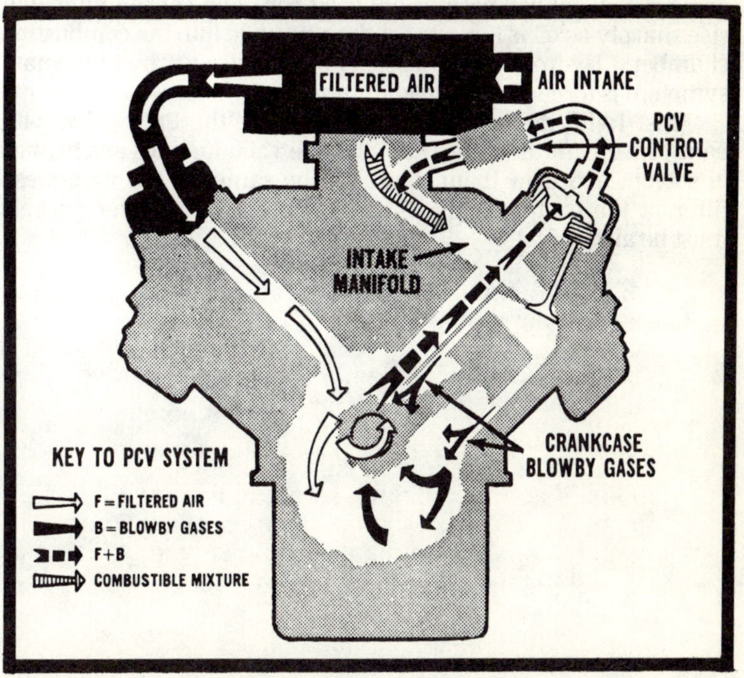

Fig. 2-18. How the positive crankcase ventilation system controls crank case blowby. (Courtesy Chevrolet Motor Div., General Motors Corp.)

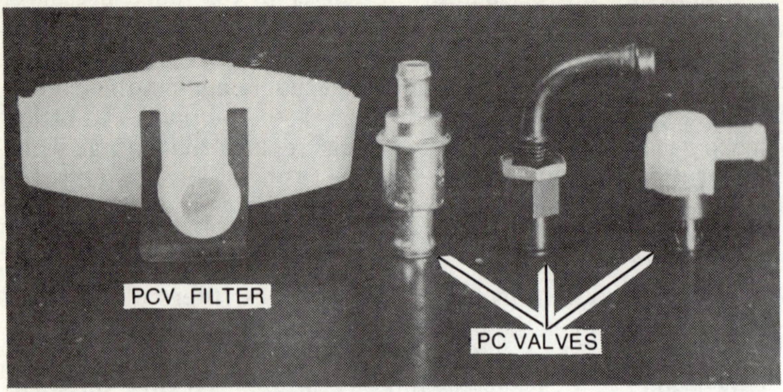

Fig. 2-19. Initial PCV system components.

technology and in some ways the most successful. Without any mileage or driveability penalties, PCV reduced hydrocarbon emissions by 25%.

Several variants have been used, but all domestic cars employ a vacuum-controlled PCV valve to regulate the amount of crankcase fumes sent to the intake manifold (Fig. 2-19).

Should the system clog, the crankcase will become pressurized. Oil will ooze from every pore and consumption will rise sharply as oil is forced past the rings and into the combustion chambers. Hydrocarbon emissions will increase. The immediate symptom is a rough, erratic idle.

Test (with the proper gage) or replace the PCV valve with each tuneup. At the same time soak the vacuum lines and fittings in solvent and blow them clear. Do the same for the wire-mesh filter at the oil filler cap. Crankcase and engine-cover gaskets must be airtight.

Chapter 3

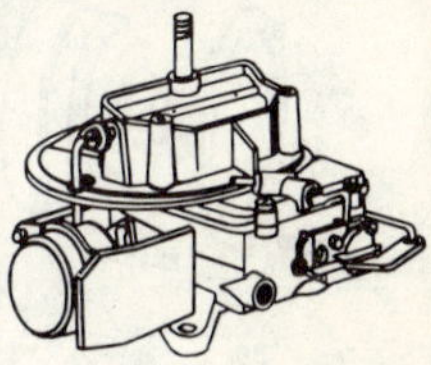

Carter

Carter is one of the oldest and most respected names in carburetors. They have at one time or another supplied carburetors to most of the major manufacturers and were one of the pioneers in four-barrel technology. Their products are distinguished by functional simplicity and are generally easy to repair. However, a word of caution—do not interchange parts between barrels on the AFB models. Keep them separate and assemble as they were originally installed.

CARTER ONE-BARREL CARBURETOR (MODEL BBS)

Please refer to Chapter 1 for a brief description and discussion of carburetor cleaners and the container needed to make up the bath to be used for this task. Tools are listed in Table 3-1.

Table 3-1. Tools Required for Carter Carburetor Overhaul.

Screwdriver with a broad blade (approx. 7/16 in.)
Screwdriver with a narrow blade (approx. 3/16 in.).
Needle-nosed pliers.
9/16 in. open-or box-end wrench.
Adjustable wrench with at least a 9/16 in. span.
Open-end, box-end, or socket wrench of the proper size to remove the nuts or bolts that hold your carburetor to the manifold.
Note: Most air cleaners can be removed without tools or with only a pair of pliers.

Preliminary Instructions

Remove air cleaner.

Remove all dirt and grime from exterior of carburetor while carburetor is still mounted on engine; use kerosene and a paintbrush if necessary, but **do not use gasoline**.

Remove carburetor from engine.

For information on carburetor accessories, refer to Chapter 2. If you are not sure you can remember where and how the lines and linkages connect to the carburetor, make notes and sketches in the margin of this book as you remove the carburetor. Figure 3-1, an exploded view, is included to help you identify the various parts of your carburetor.

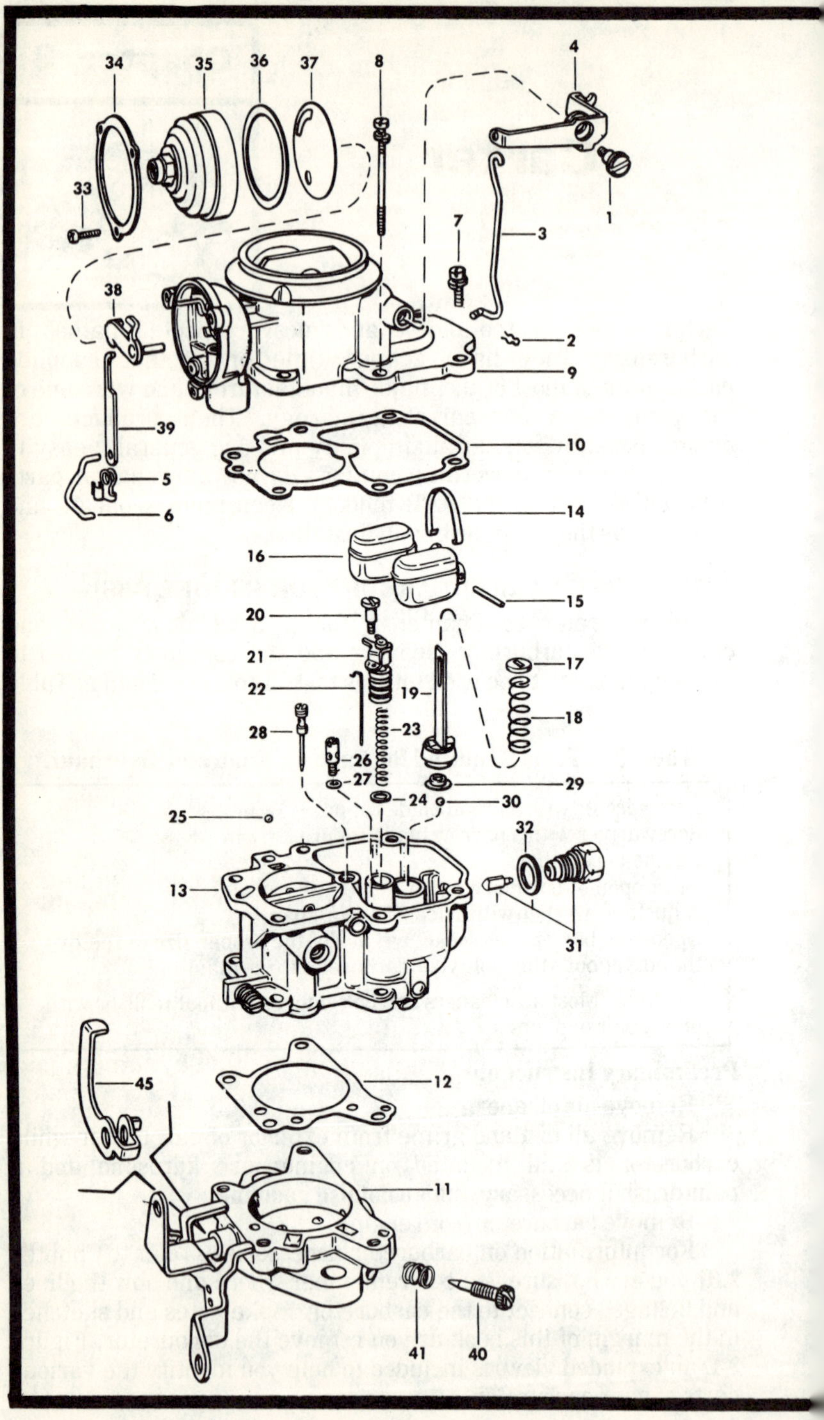
34
35
36
37
8
4
1
33
7
3
38
2
9
39
10
5
6
14
16
15
20
21
17
22
19
18
28
23
26
27
29
24
30
25
32
13
31
45
12
11
41
40

Ref. No.	NOMENCLATURE
1	Screw - Pump arm
2	Spring - Pin
3	Rod - Throttle connector
4	Arm - Pump
5	Retainer
6	Rod - Choke connector
7	Screw and Lock Washer Assembly
8	Screw and Lock Washer Assembly
9	Air Horn Assembly
10	Gasket - Body
11	Flange Assembly - Body
12	Gasket - Body flange
13	Main Body
14	Retainer - Float pin
15	Pin - Float
16	Float Assembly
17	Washer - Pump spring retainer
18	Spring - Pump
19	Piston Assembly - Pump
20	Screw - Step-up piston guide
21	Piston Assembly - Step-up
22	Rod - Step-up
23	Spring - Step-up piston
24	Gasket - Step-up piston
25	Ball - Pump discharge
26	Jet Assembly - Main metering
27	Gasket - Main metering jet
28	Jet - Low speed
29	Retainer - Pump intake ball
30	Ball - Pump intake
31	Needle and Seat Assembly
32	Gasket - Needle seat
33	Screw
34	Retainer - Coil housing
35	Coil and Housing Assembly - Thermostatic
36	Gasket - Coil housing
37	Plate - Choke baffle
38	Lever Assembly - Choke trip
39	Link - Fast idle
40	Screw - Idle adjusting
41	Spring - Idle adjusting screw
42	Bracket - Dashpot
43	Nut
44	Dashpot Assembly
45	Lever - Dashpot (with Powerflyte)
46	Bracket - Kickdown switch
47	Nut
48	Switch Assembly - Kickdown

NOTE: IDLE ADJUSTING SCREW (40) IS NOT REMOVABLE FROM CARBURETORS USED ON ENGINES WITH CLEAN AIR PACKAGE.

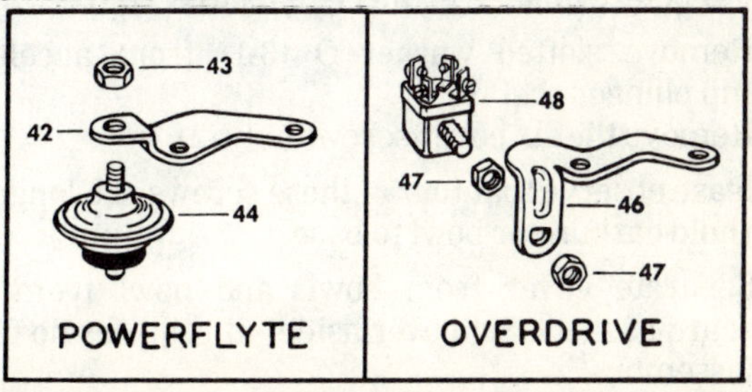

Fig. 3-1. Exploded view of Carter one-barrel carburetor, Model BBS.

Fig. 3-2. External linkage and fittings.

When carburetor is removed from engine, follow tuneup steps listed below.

To Disassemble

Step 1. Remove clip A (Fig. 3-1) and remove linkage B.

Step 2. Remove screw C (Fig. 3-2) and remove accelerator arm D.

Step 3. Remove clip E (3-2) and remove linkage F. (Parenthesized numerals refer to figures.)

Note: On some carburetors it may be necessary to remove cover screws G before linkage F can be maneuvered free.

Step 4. Remove slotted washer O (3-1) from accelerator pump plunger shaft I.

Step 5. Remove the six cover screws G (3-2).

Note: Please observe that four of these screws are longer and also serve to hold carburetor bowl to base.

Step 6. Separate cover from bowl, and bowl from base. Discard gaskets. Set cover aside—it requires no further disassembly.

Step 7. Remove idle-mixture screw H (3-2) and spring H1. Set base aside—it requires no further disassembly.

Fig. 3-3. Accelerator pump.

Step 8. Lift accelerator plunger I (3-3) with spring I1 and step-shaped washer I2 from well. Discard spring and washer.

Step 9. Remove *U*-shaped spring clip J (3-4).

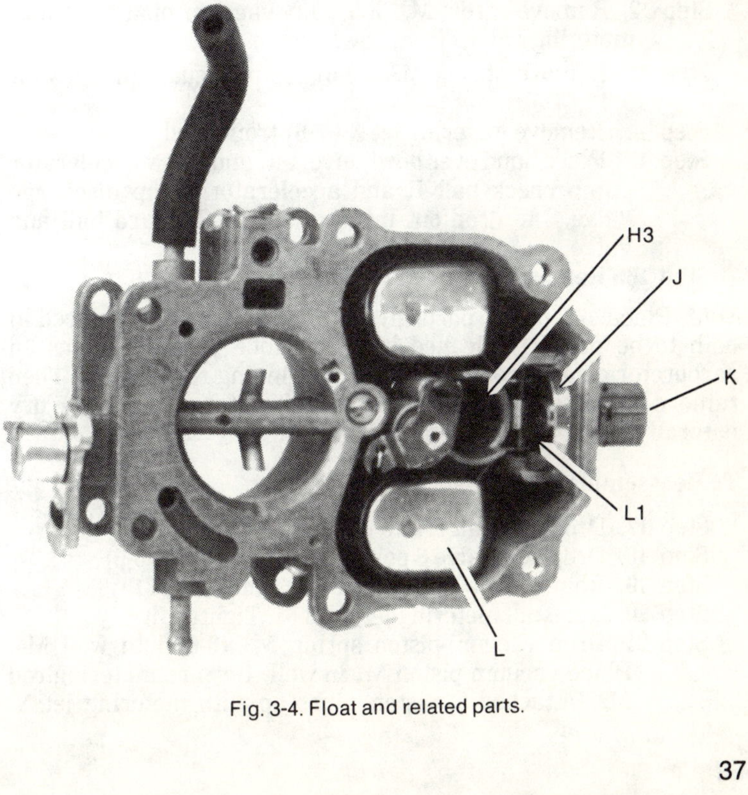

Fig. 3-4. Float and related parts.

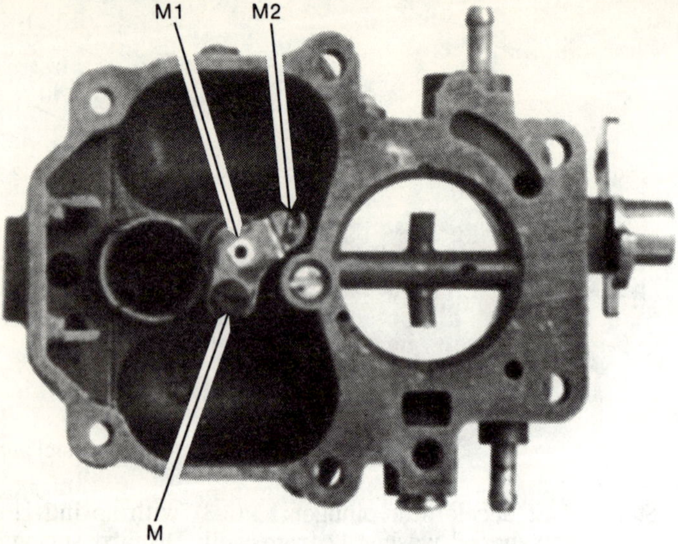

Fig. 3-5. Vacuum piston and metering rod.

Step 10. Remove needle and seat assembly K (3-4).
Step 11. Lift float L (3-4) and pin L1 from bowl.
Step 12. Remove screw M (3-5); lift vacuum piston M1 and metering rod M2 from bowl.
Step 13. Remove spring M3 (3-6) from vacuum-piston well M4.
Step 14. Remove metering jet N (3-6) from bowl.
Step 15. Place hand over bowl, invert it, and allow accelerator pump check ball R and accelerator pump discharge valve S to drop out into your hand. Discard ball and valve.
Step 16. Remove low-speed jet T (3-6).

Note: Place all metal parts in bath. Allow all parts placed in bath to be properly cleaned in accordance with directions on carburetor-cleaner container before beginning reassembly. Then remove all parts from cleaner, rinse with water, and allow to dry naturally. Do not use compressed air.

To Reassemble

Step 17. Drop new ball R into well H3 (3-6).
Step 18. Drop new valve S point first into well S1 (3-6).
Step 19. Reinstall low-speed jet T (3-6) into well T1.
Step 20. Reinstall metering jet N (3-6). Tighten firmly.
Step 21. Drop vacuum-piston spring M3 (3-6) into well M4. Place vacuum piston M1 in well. Be sure metering rod M2 (attached to piston) lines up with metering jet N.

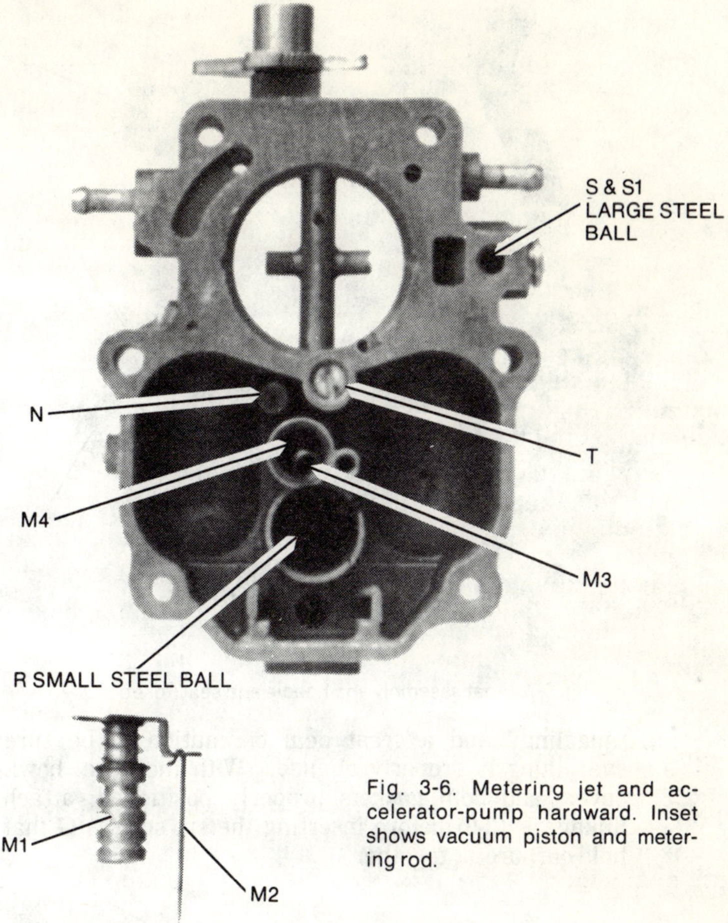

Fig. 3-6. Metering jet and accelerator-pump hardward. Inset shows vacuum piston and metering rod.

Hold down assembly with finger while replacing step-shaped screw M.

Step 22. Replace float L (3-7) and pin L1 into bowl.

Step 23. Use new gasket. Install new needle-and-seat assembly K (3-7). Tighten firmly.

Step 24. Make float-level adjustment in accordance with instruction sheet included in your tuneup parts kit.

Step 25. Install *U*-shaped spring clip J (3-7).

Step 26. Install new accelerator pump plunger I (3-8) into well. Place spring I1 and step-shaped washer I2 on plunger shaft.

Step 27. Install idle-mixture spring H1 (3-7) and screw H.

Step 28. Install new gaskets between base and bowl and between bowl and cover. (This operation requires some

Fig. 3-7. Float assembly and needle and seat (inset).

"juggling" and a great deal of caution to be sure everything is properly aligned.) With the base, bowl, cover, and both gaskets properly positioned, attach linkage F (3-9) before inserting the six screws G that hold carburetor together.

Fig. 3-8. Accelerator pump.

Fig. 3-9. External linkage and fittings.

Step 29. Insert in the proper places the four long screws G (3-9) and two short screws that hold carburetor together. Tighten firmly.

Step 30. Replace spring clip E (3-9) on linkage.

Step 31. Place slotted washer O (3-9) over accelerator pump plunger shaft I.

Step 32. Hook the end of pump-actuating arm D (3-9) into pump plunger shaft I. Secure in place with large step screw C. Tighten screw firmly.

Step 33. Hook linkage B (3-9) into arm D and attach at B1 with spring clip A. (Place in center hole B1.)

Step 34. Mount carburetor on engine. Be sure all lines and linkages are properly connected.

Step 35. Start engine. Let engine warm up until choke valve is fully open.

Step 36. Adjust idle to desired speed with idle speed adjustment screw Q (3-10).

Step 37. Adjust idle mixture; turn screw H (3-10) clockwise until engine begins to run rough; then turn screw counterclockwise to obtain smooth operation. Mount air cleaner on carburetor. Be certain element is clean.

Fig. 3-10. Idle-speed and mixture screws.

CARTER ONE-BARREL CARBURETOR (MODEL YF)

Please refer to Chapter 1 for a brief description and discussion of carburetor cleaners and the container needed to make up the bath to be used for this task. Tools are listed in Table 3-1.

Preliminary Instructions

Remove air cleaner.

Remove all dirt and grime from exterior of carburetor while carburetor is still mounted on engine; use kerosene and a paintbrush if necessary, but **do not use gasoline**.

Remove carburetor from engine.

For information on carburetor accessories, refer to Chapter 2. If you are not sure you can remember where and how the lines and linkages connect, make notes and sketches in the margin of this book as you remove the carburetor. Figure 3-11, an exploded view, is included to help you identify the various parts of your carburetor.

When carburetor is removed from engine, follow tuneup steps listed below.

To Disassemble

Step 1. Remove three screws A1 (3-12), retainers A2 and choke housing A. Do *not* place in bath.

Step 2. Remove clip B1 (3-12) and disconnect choke linkage. Place linkage in bath.

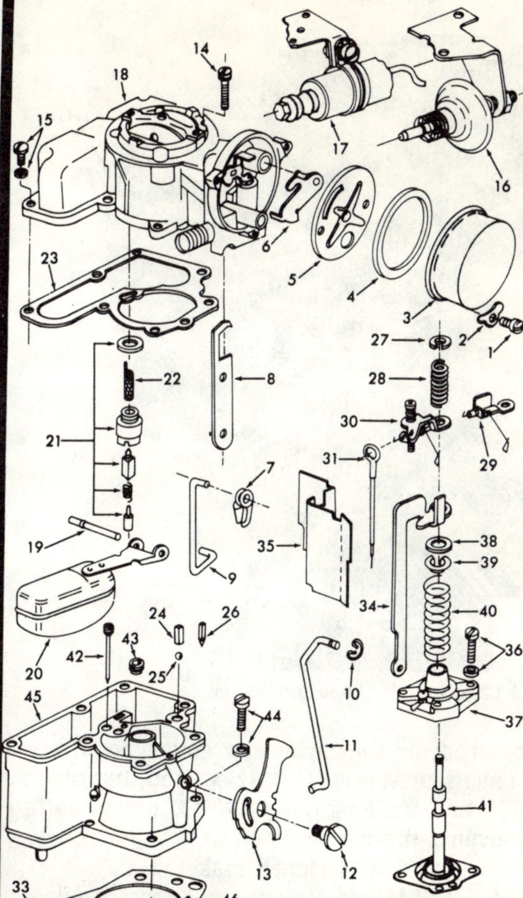

NOMENCLATURE

REF. NO.

1. SCREW(3)-CHOKE COVER CLAMP
2. CLAMP(3)-CHOKE COVER
3. CHOKE COVER & SPRING
4. GASKET - CHOKE COVER
5. BAFFLE PLATE - CHOKE
6. TRIP LEVER - CHOKE (EARLY MODELS)
7. RETAINER-FAST IDLE LINK (EARLY MODELS)
8. LINK-FAST IDLE (EARLY MODELS)
9. ROD-FAST IDLE (EARLY MODELS)
10. RETAINER-FAST IDLE ROD
11. ROD - FAST IDLE
12. SCREW - FAST IDLE CAM
13. CAM - FAST IDLE
14. SCREW & LOCKWASHER (3) LONG - BOWL COVER
15. SCREW & LOCKWASHER (4) SHORT - BOWL COVER
16. DASHPOT & BRACKET - (WHEN USED)
17. SOLONOID & BRACKET - (WHEN USED)
18. BOWL COVER ASSEMBLY
19. PIN - FLOAT
20. FLOAT & LEVER ASSEMBLY
21. NEEDLE, SEAT AND GASKET ASSEMBLY
22. SCREEN - NEEDLE SEAT (SOME MODELS)
23. GASKET - BOWL COVER
24. WEIGHT - CHECK BALL
25. BALL - PUMP CHECK
26. NEEDLE - PUMP CHECK (SOME MODELS)
27. RETAINER - UPPER PUMP SPRING
28. SPRING - UPPER PUMP
29. ARM ASSY. - METERING ROD (EARLY MODELS)
30. ARM & ADJ. SCREW ASSY.- METERING ROD
31. ROD - METERING
32. ARM - THROTTLE SHAFT
33. LINK - PUMP CONNECTOR
34. LINK - PUMP LIFTER
35. BAFFLE PLATE - FUEL BOWL (SOME MODELS)
36. SCREW & LOCKWASHER (4)- PUMP HOUSING
37. PUMP HOUSING ASSEMBLY
38. SPACER - PUMP STEM (SOME MODELS)
39. RETAINER - PUMP SPRING
40. SPRING - PUMP
41. PUMP DIAPHRAGM ASSEMBLY
42. JET - LOW SPEED
43. JET - METERING ROD
44. SCREW & LOCKWASHER (3)- THROTTLE BODY
45. BOWL ASSEMBLY
46. GASKET - THROTTLE BODY
47. CAP - IDLE LIMITER
48. NEEDLE - IDLE ADJUSTING
49. SPRING - IDLE NEEDLE
50. THROTTLE BODY ASSEMBLY

Fig. 3-11. Exploded view of Carter one-barrel carburetor, Model YF.

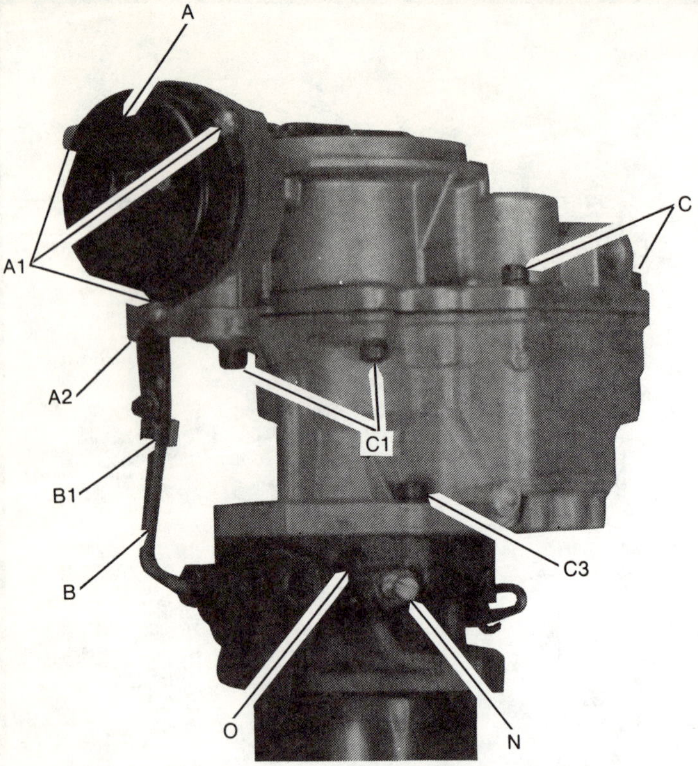

Fig. 3-12. External linkage and fittings.

Step 3. Seven screws secure the carburetor cover, bowl, and base into a unit: four screws C (3-12) on top, two short ones C1 (3-12), and one long one C2 (3-13) underneath. Remove all seven of these screws.

Step 4. Lift bowl cover off bowl. Discard gasket.

Step 5. Slip float pin D1 (3-14) out of float hinge; remove float D, but do *not* place float in bath.

Step 6. Remove needle valve E (3-14) from well. Discard this valve.

Step 7. Use broad-blade screwdriver; remove needle-valve seat E1 (3-14) from well. Discard seat.

Step 8. Place bowl cover in bath.

Step 9. Loosen screw F1 (3-13); remove bracket F2 and attached linkage from throttleshaft F. Place bracket and linkage in bath.

Step 10. From top of accelerator pump plunger assembly G (3-15), remove clip G1 while holding coil spring G2 to keep it from popping off. Remove spring and place it in bath.

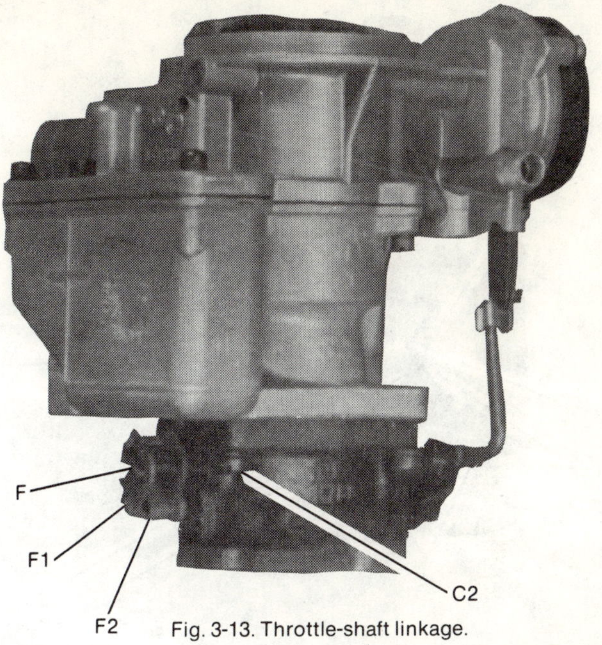

Fig. 3-13. Throttle-shaft linkage.

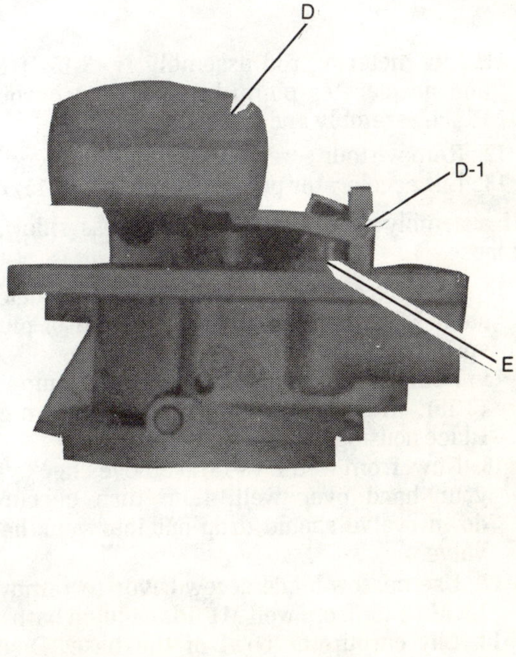

Fig. 3-14. Float and needle and seat.

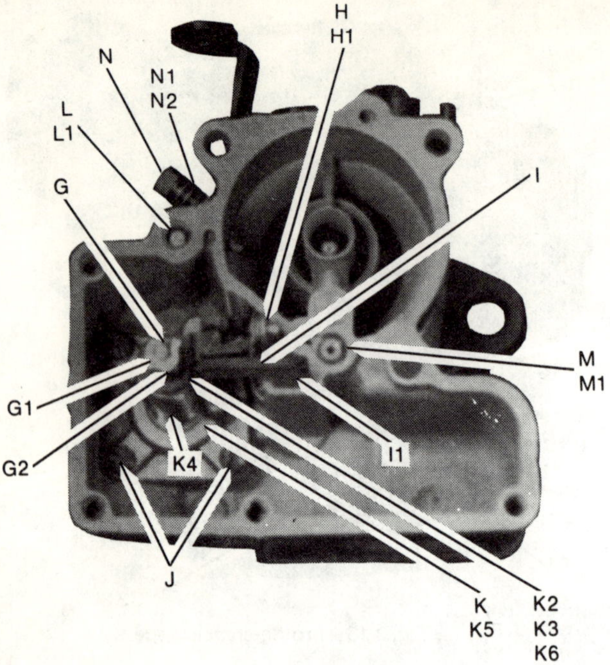

Fig. 3-15. Bowl internals.

Step 11. Lift metering rod assembly H (3-15) from well H1, and accelerator pump lever I from carburetor bowl. Place assembly and lever into bath.

Step 12. Remove four screws J (3-15).

Step 13. Lift accelerator pump assembly K (3-15) out of bowl.

Note: If assembly seems to be stuck, press sideways on it to break it loose.

Step 14. Remove clip K2 (3-15) from top of accelerator pump shaft K3, then slip coil spring K4 off of plunger. Place spring in bath.

Step 15. Remove and discard accelerator pump plunger K3 (3-15), with diaphragm attached, from metal housing. Place housing in bath.

Step 16. Now, from well L (3-15), remove check valve L1: cup your hand over well, then turn carburetor upside down—valve should drop out into your hand. Discard valve.

Step 17. Use narrow-blade screwdriver to remove low-speed jet M (3-15) from well M1. Place jet in bath.

Step 18. Lift carburetor bowl of the base. Discard gasket. Place bowl in bath.

Step 19. From well N2 (3-15) in base of carburetor, remove idle mixture adjustment screw N and spring N1. Place screw and spring in bath.

Step 20. Place carburetor base in bath.

You have completed the carburetor disassembly. Allow all parts placed in bath to be properly cleaned in accordance with directions on cleaner container before beginning reassembly. The remove all parts from cleaner, rinse them with water, and allow them to dry naturally. Do not use compressed air.

To Reassemble

Step 21. Place spring N1 (3-15) over idle-mixture adjustment screw N and insert screw into well N2 in carburetor base. Tighten screw until it seats lightly, then back it out 1½ turns.

Step 22. Use new gasket. Place carburetor bowl into position upon base. Insert two screws C3 (3-16) and tighten them.

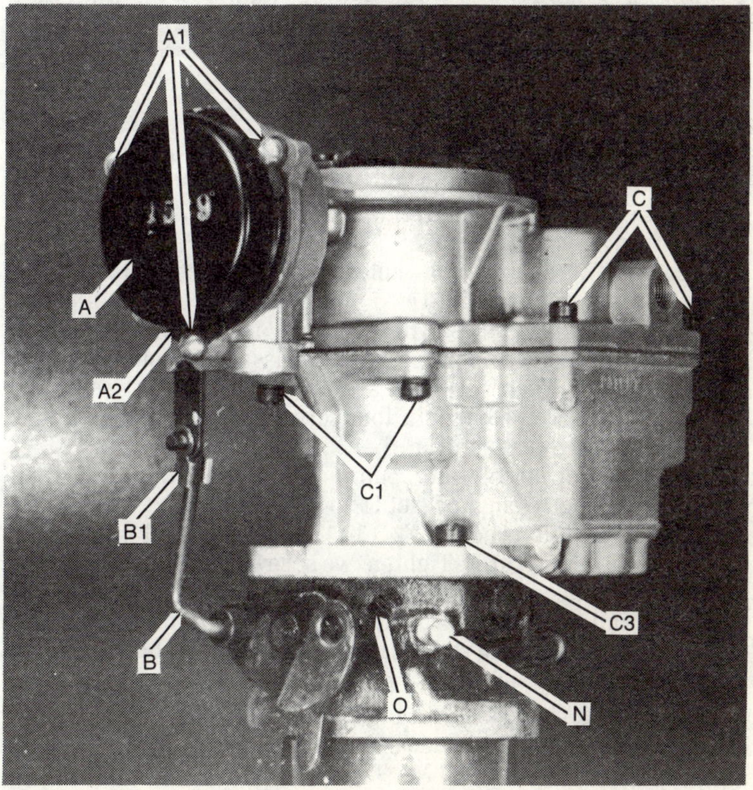

Fig. 3-16. Automatic choke, cover-to-bowl screws, and mixture-adjustment screw.

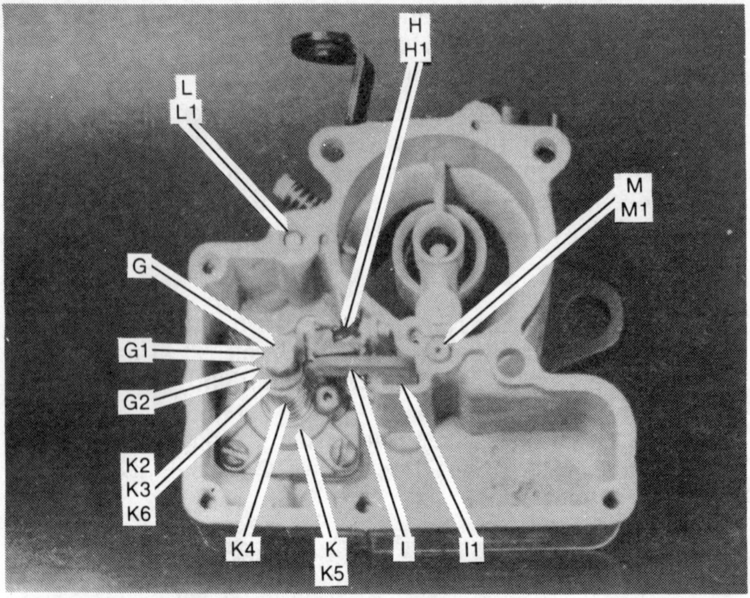

Fig. 3-17. Bowl internals.

Step 23. Place new accelerator pump shaft K3 (3-17), with diaphragm K6 attached, into metal housing K5.

Step 24. Slip coil spring K4 (3-17) over shaft K3 and secure it in place with clip K2.

Step 25. There are five holes in the accelerator pump diaphragm K6 (3-17). Four are equally spaced—these are screw holes. The fifth hole is a fuel passage—be sure it lines up properly with passage in housing and in bowl. When passages are properly aligned, insert the four screws J into the four screw holes in the diaphragm and tighten firmly.

Step 26. Drop accelerator pump lever I (3-17) into position through elongated slot I1.

Step 27. Install bracket F2 (3-18) and attached linkage to throttle shaft F. Tighten two screws F1.

Step 28. Place metering rod assembly H (3-17) into well H1 and in position around accelerator pump shaft K3. Be sure metering rod fits into jet at bottom of bowl.

Step 29. Place coil spring G2 (3-17) over accelerator pump shaft G and fasten it in place with clip G1.

Step 30. Use narrow-blade screwdriver to install low-speed jet M (3-17) into well M1. Tighten firmly.

Step 31. Drop new needle check valve L1 (3-17) point first into well L.

Fig. 3-18. Throttle-shaft linkage.

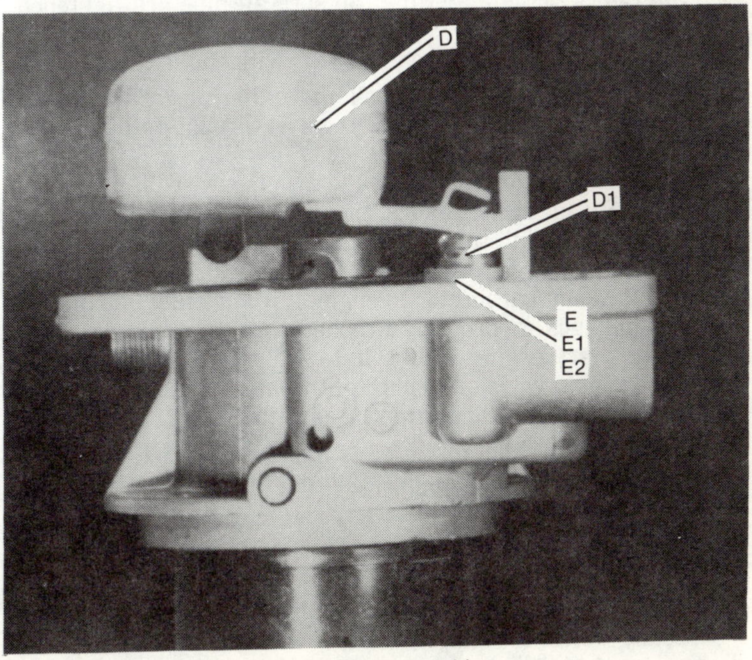

Fig. 3-19. Float assembly.

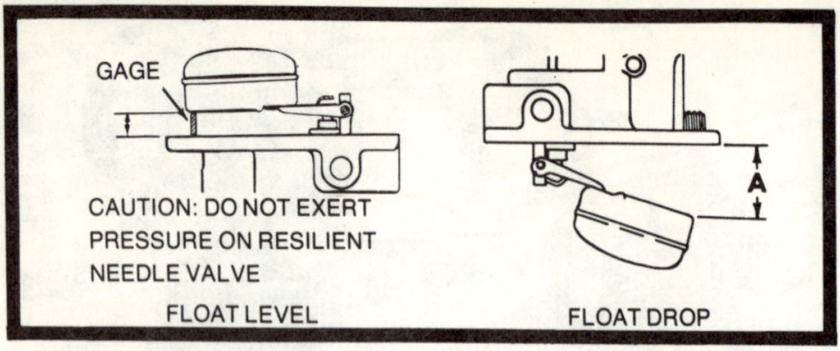

Fig. 3-20. Float adjustment. Measurement is without gasket for this model.

Step 32. Install new valve seat E1 (3-19) and gasket into well E2. Tighten firmly.

Step 33. Drop new needle valve E (3-19) point first into seat E1, which is in well E2.

Step 34. Install float D (3-19); insert hinge pin D1 into hinge—be sure pinhead is toward the outside.

Step 35. Adjust float level. Refer to instruction sheet in parts kit for level setting, and to Fig. 3-20. *Be Accurate*.

Step 36. Place new gasket on bowl and place bowl cover into position on bowl. Insert four screws C (3-21) and screws C2. Tighten these seven screws, and tighten screws C3 (3-21) at the ame time. Be sure all nine are tightened firmly.

Step 37. Attach choke linkage B (3-21) with clip B1.

Step 38. Use new gasket. Install choke-thermostat housing A (3-21): insert the three screws A1 through retainers A2, but do not tighten screws yet.

Step 39. Rotate choke housing A (3-21) clockwise until you feel it engage choke shaft; then turn it until mark on housing lines up with the center mark on the metal base. Now, tighten three screws A1 (3-21).

Step 40. Mount carburetor to engine. Be sure fuel line and all linkages are properly attached.

Step 41. Start engine. Allow it to warm up until butterfly (choke) valve in carburetor throat is fully open.

Step 42. Adjust idle to desired speed with screw O (3-21).

Step 43. Adjust idle mixture with screw N (3-21): turn it clockwise until engine falters, then back it out until engine runs smoothly.

Replace air cleaner on carburetor. Be sure element is clean.

CARTER TWO-BARREL CARBURETOR (MODEL WGD)

Please refer to Chapter 1 for a brief description and discussion of carburetor cleaners and the container needed to

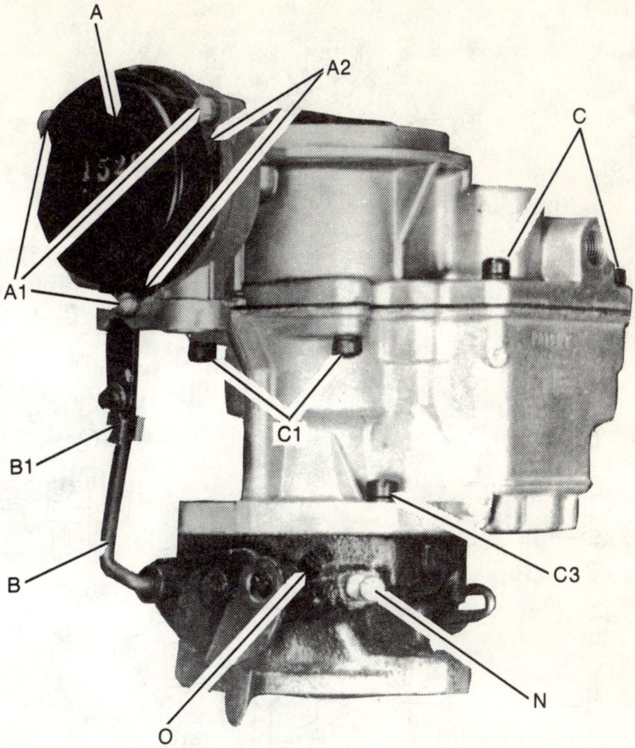

Fig. 3-21. External fasteners and fittings.

make up the bath to be used for this task. Tools are listed in Table 3-1.

Preliminary Instructions

Remove air cleaner.

Remove all dirt and grime from exterior of carburetor while carburetor is still mounted on engine; use kerosene and a paintbrush if necessary, but **do not use gasoline**.

Remove carburetor from engine.

For information on carburetor accessories, refer to Chapter 2. If you are not sure you can remember where and how the lines and linkages connect, make notes and sketches in the margin of this book as you remove the carburetor. Figure 3-22, an exploded view, is included to help you identify the various parts of your carburetor.

When carburetor is removed from engine, follow tuneup steps listed on page 53.

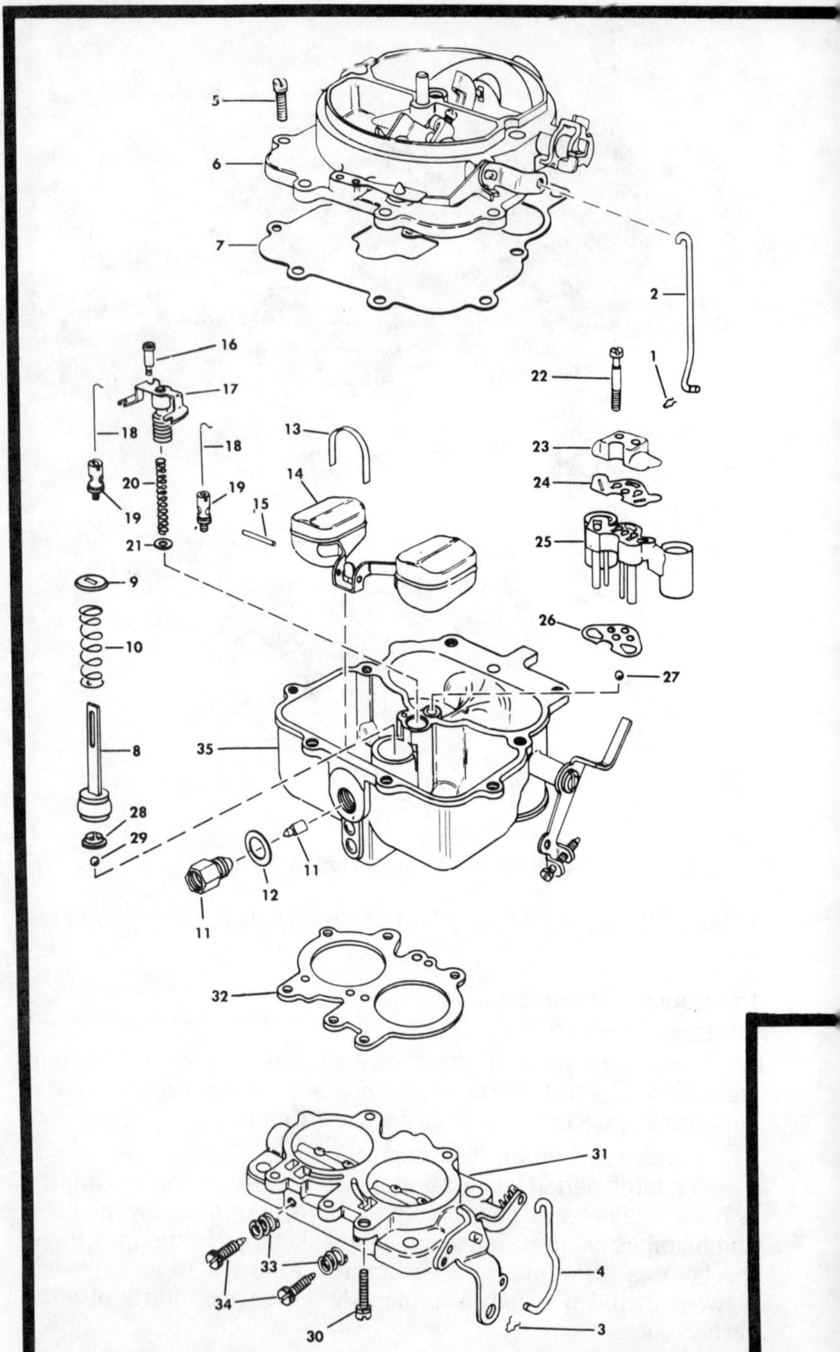
1
2
3
4
5
6
7
8
9
10
11
11
12
13
14
15
16
17
18
18
19
19
20
21
22
23
24
25
26
27
28
29
30
31
32
33
33
34
34
35

REF. NO.	NOMENCLATURE
1	Pin spring
2	Throttle-connector rod
3	Pin spring
4	Fast-idle connector rod
5	Screw-and-lockwasher assembly
6	Air-horn assembly
7	Body gasket
8	Pump-plunger assembly
9	Pump-spring retainer
10	Pump spring
11	Needle-and-seat assembly
12	Needle-seat gasket
13	Float-pin retainer
14	Float assembly
15	Float pin
16	Step-up piston screw
17	Step-up piston assembly
18	Step-up rods
19	Main-metering jets
20	Step-up piston spring
21	Step-up piston gasket
22	Venturi-cluster screw
23	Venturi cover
24	Venturi-cover gasket
25	Venturi-cluster assembly
26	Venturi-housing gasket
27	Pump-discharge check ball
28	Intake check-ball retainer
29	Pump intake check-ball
30	Screw-and-lockwasher assembly
31	Throttle-body assembly
32	Throttle-body gasket
33	Idle-adjusting screw springs
34	Idle-adjusting screws
35	Carburetor-body assembly

NOTE: IDLE ADJUSTING SCREWS (34) ARE NOT REMOVABLE FROM CARBURETORS USED ON ENGINES WITH CLEAN AIR PACKAGE. (PRIOR TO 1968) 1968-69 MODELS HAVE A SINGLE IDLE MIXTURE SCREW WITH A LEFT HAND THREAD. 1968-69 MODELS HAVE SPECIAL ADJUSTMENT SCREWS PLUGGED AND FACTORY SEALED. DO NOT REMOVE OR ADJUST THESE SCREWS.

Fig. 3-22. Exploded view of the Carter two-barrel carburetor, Model WGD.

To Disassemble

Step 1. Remove hairpin retainer A (3-23); disconnect linkage B at A1 only.

Step 2. Remove spring retainer C (3-23); disconnect linkage D at C1 only.

Step 3. Remove hairpin retainer E (3-23); disconnect linkage E1 at both ends and set it aside.

Step 4. Remove two screws F (3-23); lift off cover FF. Discard gasket.

Step 5. Examine carefully the metering rod assembly G (3-24), paying special attention to how rods are held in

Fig. 3-23. Top view.

place by a tiny spring H. Push metering rods G1 off ends of activating rod G2; then pull them straight out, being very careful not to bend them. (If a metering rod is bent it should be replaced. Metering rods are *not* included in parts kit.) Put metering rods aside.

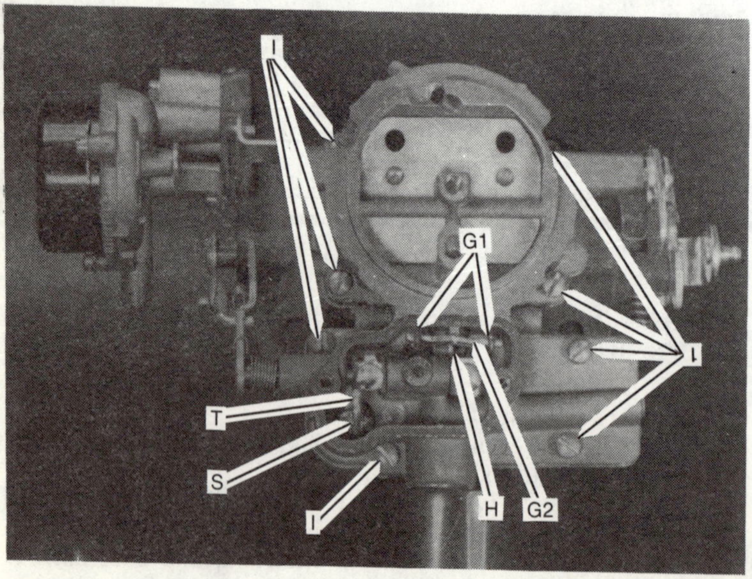

Fig. 3-24. Top view prior to removing bowl cover.

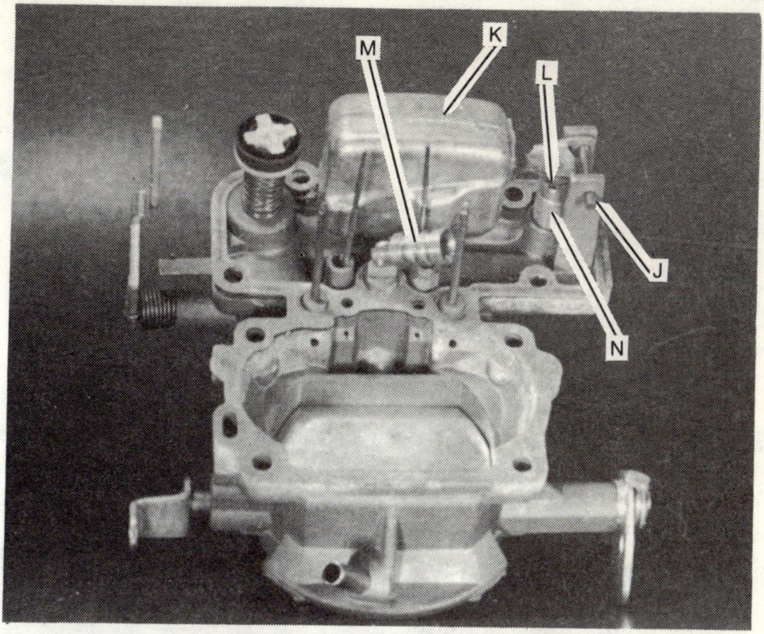

Fig. 3-25. Float, vacuum piston, and related parts.

Step 6. Remove the eight bowl-cover screws I (3-24); lift off bowl cover and turn it upside down.

Step 7. Slide out float pin J (3-25); lift off float assembly K and set it aside.

Step 8. Lift out and discard needle valve L (3-25).

Step 9. Tilt vacuum piston M (3-25) to left; lift it out and set it aside.

Step 10. Remove needle-valve seat N (3-25)—use broad-blade screwdriver. Discard seat.

Step 11. Remove three screws O (3-26) and retainer ring P; then remove Bakelite choke cover Q. Set it aside.

Step 12. Remove and discard the bowl-cover gasket.

Step 13. Push down on accelerator pump assembly R (3-26), and remove hairpin retainer S (3-26); pull pump linkage T gently to right to free it, then remove and discard accelerator pump assembly, but keep spring U (3-26). Place spring in bath.

Step 14. Place bowl cover assembly in bath.

Step 15. Remove small spring W from well W1 (3-27); set spring aside.

Step 16. Remove the two screws X (3-27); lift screws and accelerator-discharge nozzle Y off and set aside. Discard gasket.

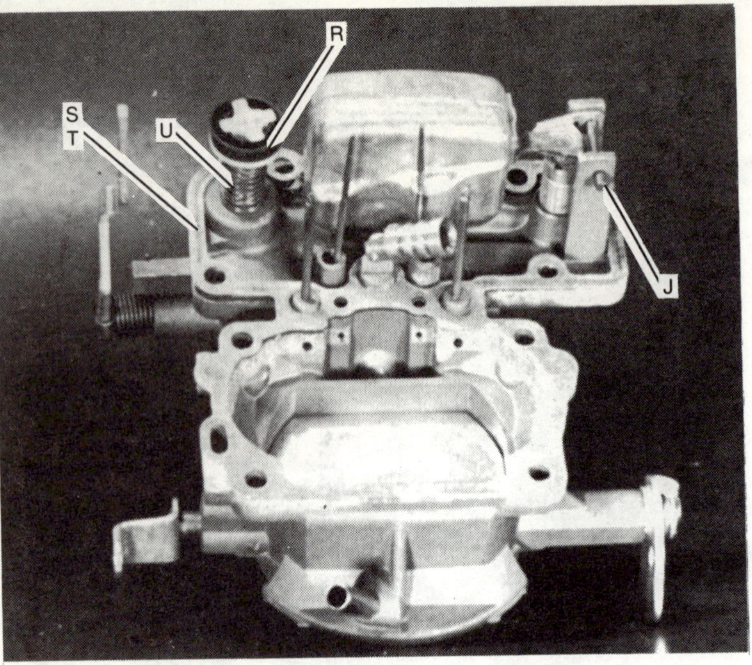

Fig. 3-26. Accelerator-pump and related parts.

Step 17. Cup your hand over well at Z (3-27), turn carburetor assembly over—a check valve or ball Z1 should fall out of well into your hand. If it's a valve, set it aside; if it's a ball, discard it.

Fig. 3-27. Choke and pump-discharge parts.

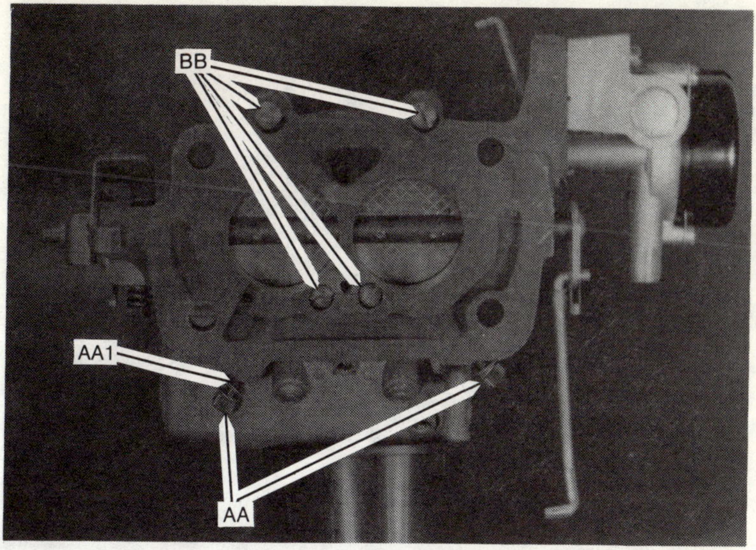

Fig. 3-28. Throttle-plate assembly.

Step 18. Turn carburetor over. Remove idle-mixture screws AA (3-28) and springs AA1. Set them aside.

Step 19. Remove the four screws BB (3-28) that hold base to carburetor bowl. Separate base from bowl. Discard gasket. Place base and bowl in bath.

Note: All metal parts indicated should now be placed in bath. Allow parts placed in bath to be properly cleaned in accordance with directions on carburetor-cleaner container before beginning reassembly. The remove all parts from cleaner, rinse with water, and allow to dry naturally. Do not use compressed air.

To Reassemble

Step 20. Use new gasket; assemble base to bowl; tighten four screws BB (3-28) evenly and firmly.

Step 21. Replace springs AA1 on idle-mixture screws AA (3-28); insert screws all the way, then back them out 1½ turns.

Step 22. Turn carburetor right side up. Drop check valve Z1 (3-29) or new ball into passage Z.

Step 23. Use new gasket; replace accelerator discharge nozzle assembly Y with screws X (3-29). Tighten screws.

Step 24. Replace small spring 23 (3-29).

Step 25. Be sure metering jets CC (3-29) are tight.

Step 26. Blow through both balance tubes DD (3-30) to be sure they are open.

Fig. 3-29. Accelerator-pump and related parts.

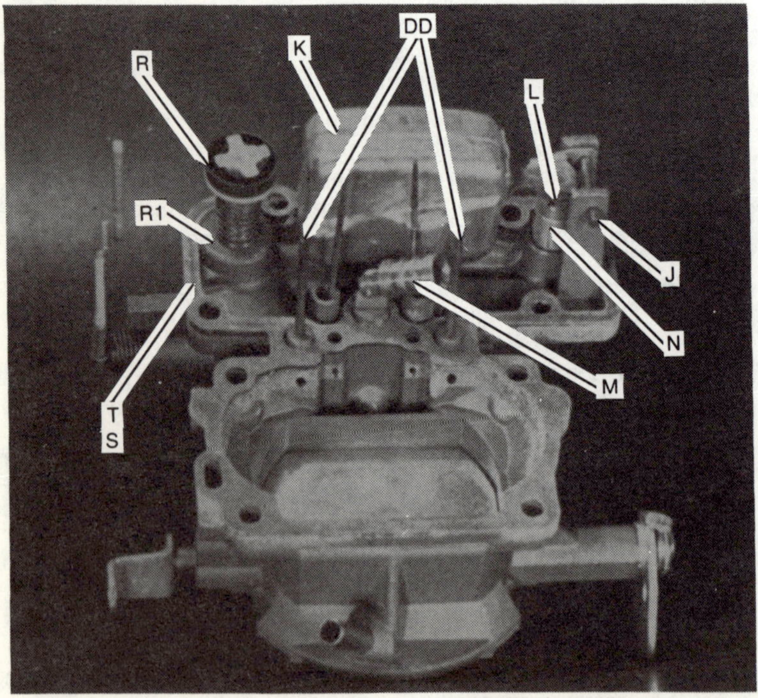

Fig. 3-30. Pump and float assembly.

Step 27. Place spring U (3-30) over shaft of new accelerator pump assembly R; insert shaft into hole R1 in bowl cover; push down on assembly, attach linkage T, and fasten in place with hairpin retainer S.

Step 28. Use new gasket; install new needle-valve seat N (3-30). Tighten with broad-blade screwdriver.

Step 29. Place a new gasket on bowl cover.

Step 30. Replace vacuum piston M (3-30).

Step 31. Drop new needle valve L (3-30) into seat N.

Step 32. Replace float assembly K (3-30) and float pin J. Make float level adjustment in accordance with instruction sheet in parts kit. *Be accurate*.

Step 33. Place bowl cover in position over bowl and carefully lower it into position—be sure that nothing binds and that gasket is properly positioned. Insert the eight screws I (3-31), and tighten evenly and firmly.

Step 34. Replace metering rods G1 (3-31) carefully—don't bend them. First, hook them in the tiny spring H, then push them straight down into metering jets CC (3-29), and finally hook them over ends of activating rod G2 (3-3).

Step 35. Attach linkage E1 (3-31) and replace hairpin retainer E; position counterweight DD as shown in 3-33.

Step 36. Attach linkage D (3-31) and replace spring retainer C.

Step 37. Attach linkage B (3-31) and replace hairpin retainer A.

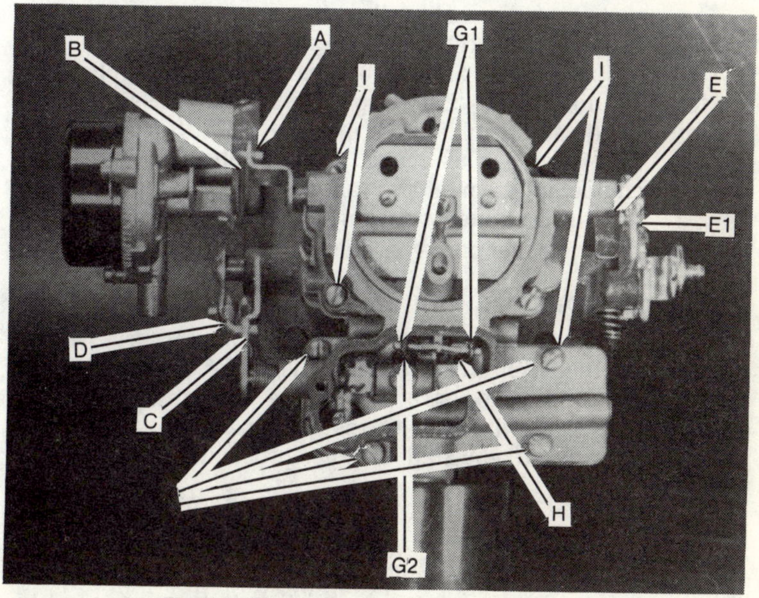

Fig. 3-31. Float-cover and linkage assembly.

Fig. 3-32. Choke cover and bowl-cover screws.

Step 38. Use new gasket. Replace cover FF (3-32) with screws F.

Step 39. Replace Bakelite choke cover Q (3-32) on choke assembly; fasten loosely with retainer P and three screws O. Turn cover clockwise as far as it will go, then return to position in which center line (if more than one) on cover aligns with center mark of raised marks on metal choke base. Choke valve in throat of carburetor should be fully open.

Step 40. Install carburetor on car.

Step 41. Start engine; allow it to warm up until choke opens.

Step 42. Adjust idle to desired speed with screw EE (3-33).

Step 43. Adjust idle mixture screws AA (3-33); turn clockwise until engine runs rough, then counterclockwise until engine idles smoothly; adjust the two idle mixture screws alternately until smoothest idle is achieved.

Replace air cleaner on carburetor. Be sure element is clean.

CARTER QUADRAJET FOUR-BARREL CARBURETOR

The Carter Quadrajet carburetor is manufactured under license from the Rochester Products Division of General Motors Corporation. Design and structure of the Carter Quadrajet is therefore identical with that of the Rochester Quadrajet. We refer you to the step-by-step tuneup procedure presented in Chapter 7.

CARTER FOUR-BARREL CARBURETOR (MODEL AFB)

Please refer to Chapter 1 for a brief description and discussion of carburetor cleaners and the container needed to

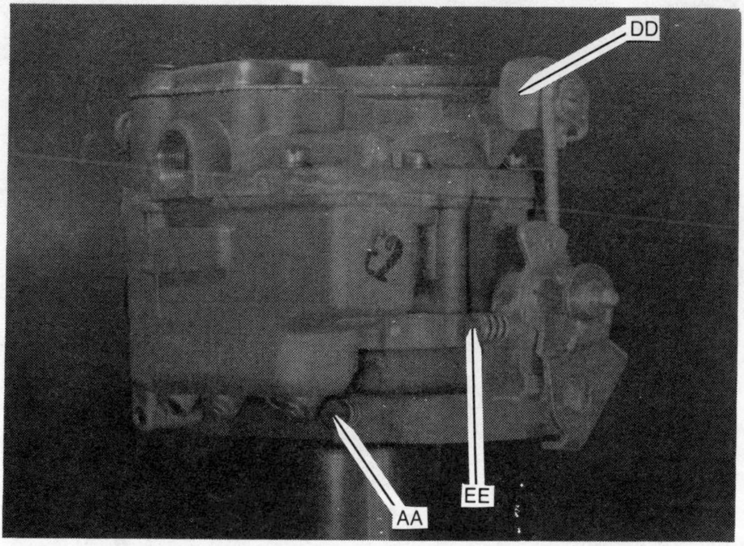

Fig. 3-33. Idle-speed and mixture screws.

make up the bath to be used for this task. Table 3-1 lists the required tools.

Preliminary Instructions

Remove air cleaner.

Remove all dirt and grime from exterior of carburetor while carburetor is still mounted on engine; use kerosene and a paintbrush if necessary, but **do not use gasoline**.

Remove carburetor from engine.

For information on carburetor accessories, refer to Chapter 2. If you are not sure you can remember where and how the lines and linkages connect, make notes and sketches in the margin of this book as you remove the carburetor. Figure 3-34, an exploded view, is included to help you identify the various parts of your carburetor.

When carburetor is removed from engine, follow tuneup steps listed below.

To Disassemble

Step 1. Remove spring clip A (3-35) and disconnect linkage B.

Step 2. Remove spring clip C (3-35) and disconnect linkage D.

Step 3. Remove screw E (3-35) and remove accelerator pump lever E1.

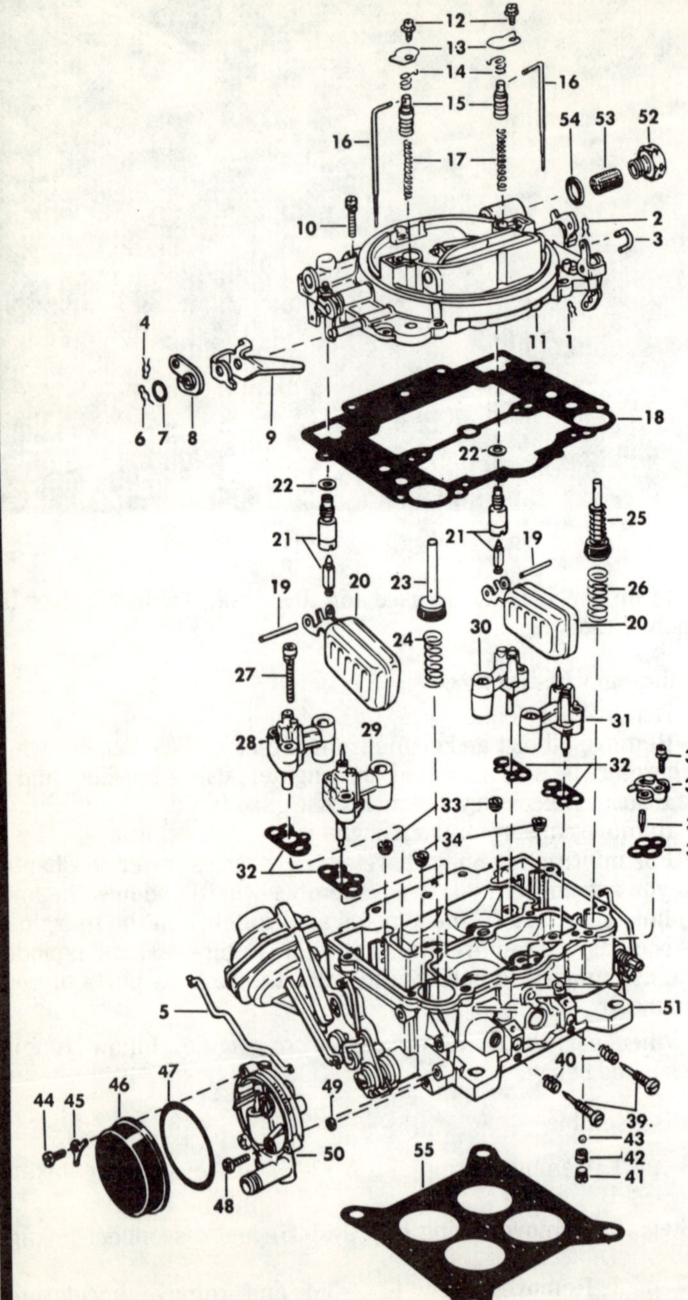
12
13
14
15
16
17
10
54
53
52
2
3
4
6
7
8
9
11
1
18
22
21
19
20
23
24
25
26
27
28
29
30
31
32
33
34
35
36
37
38
5
51
40
39
43
42
41
44
45
46
47
49
50
48
55

REF. NO.	NOMENCLATURE
1	Pin spring
2	Pin spring
3	Pump link
4	Pin spring
5	Choke connector rod
6	Pin spring
7	Dashpot lever washer
8	Outer-dashpot arm
9	Inner-dashpot arm
10	Screw and lock washer assy
11	Air horn assy
12	Screw and lock washer assys
13	Step-up piston covers
14	Step-up rod retaining springs
15	Vacuum pistons
16	Step-up rods
17	Vacuum piston springs
18	Air horn gasket
19	Float pin
20	Float assys
21	Needle and seat assys
22	Needle seat gaskets
23	Dashpot plunger assy
24	Dashpot plunger spring
25	Pump plunger assy
26	Pump return spring
27	Screw and lock washer assy
28	Secondary venturi assy — choke side
29	Primary venturi assy — choke side
30	Secondary venturi assy — pump side
31	Primary venturi assy — pump side
32	Venturi cluster gaskets
33	Secondary metering jets
34	Primary metering jets
35	Screw and lock washer assy
36	Pump jet housing
37	Pump discharge needle
38	Pump jet housing gasket
39	Idle adjusting screws
40	Idle adjusting screw springs
41	Pump intake passage plug
42	Pump intake ball seat
43	Pump intake ball
44	Screw
45	Coil housing retainer
46	Thermostatic coil and housing assy
47	Coil housing gasket
48	Screw
49	Choke piston housing assy
50	Choke piston housing gasket
51	Carburetor body assy
52	Fuel inlet fitting
53	Fuel inlet strainer
54	Fuel inlet fitting gasket
55	Flange gasket

Fig. 3-34. Exploded view of Carter four-barrel carburetor, Model AFB.

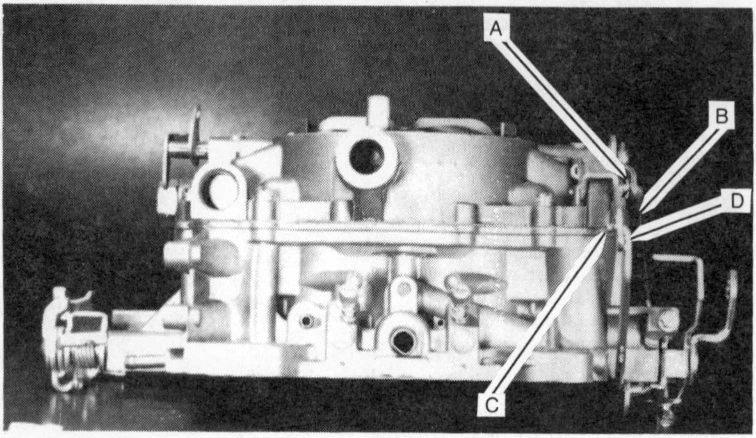

Fig. 3-35. External linkage.

Step 4. Remove screws F (3-36) and lift off retainer plates G. Lift vacuum pistons, metering rods, and return springs from wells underneath retainer plates.

Step 5. Remove eight cover screws L (3-36) and lift cover off bowl. Set cover aside.

Step 6. Slide hinge pins M out and remove both floats N (3-32).

Step 7. Remove both needle-and-seat assemblies O (3-32).

Step 8. Remove accelerator pump plunger P and return spring Q from well R (3-38).

Fig. 3-36. Bowl-cover screws.

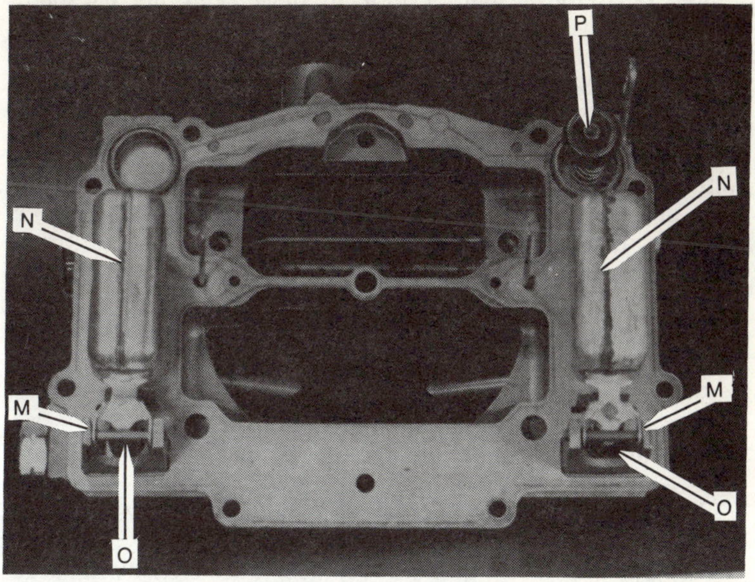

Fig. 3-37. Float assemblies.

Step 9. Remove two screws S and lift up on discharge-nozzle assembly T (3-38).

Step 10. Remove needle check valve U (3-38) from well V beneath pump discharge-nozzle assembly T.

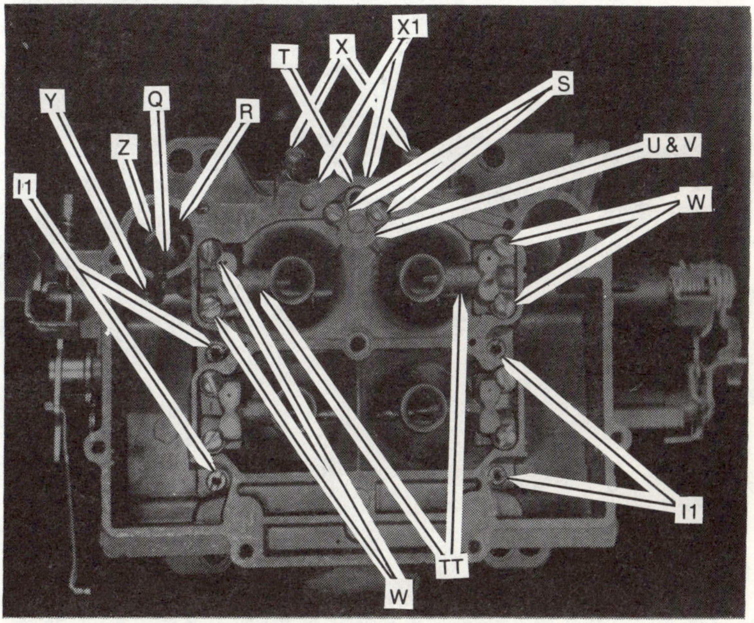

Fig. 3-38. Float-bowl hardware.

Step 11. Remove four screws W (3-38) and lift out both main-jet discharge nozzle assemblies TT.

Step 12. Remove the two idle mixture screws X and springs X1 (3-38).

Step 13. Remove accelerator pump check valve Y (3-38) from base of accelerator pump well Z.

Note: Place all metal parts in bath. Allow all parts placed in bath to be properly cleaned in accordance with directions on cleaner container before beginning reassembly. Then, remove all parts from cleaner, rinse with water, allow to dry naturally. Do not use compressed air.

To Reassemble

Step 14. Install new check valve Y (3-38).

Step 15. Tighten the four jets I1 in bottom of bowl (3-38).

Step 16. Use new gaskets; replace both main discharge nozzles TT (3-38). Tighten screws firmly.

Step 17. Drop new check valve U (point down) into well V (3-38).

Step 18. Use new gasket; replace accelerator-discharge nozzle assembly T (3-38). Tighten screws firmly.

Step 19. Replace idle mixture screws X (3-38) and springs X1. Turn screws all the way in until needle is seated, then back out each a full turn.

Step 20. Remove horseshoe-shaped retainer P1 and spring P2 from old accelerator pump plunger P (3-39); install them on new pump plunger.

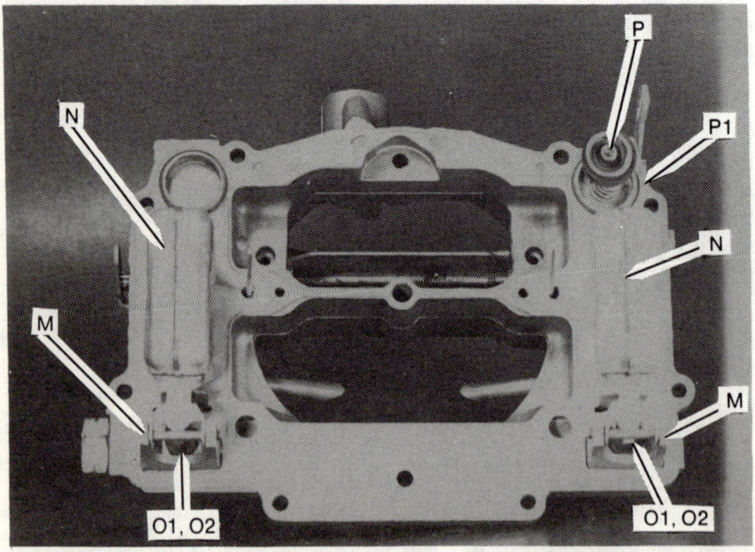

Fig. 3-39. Accelerator pump and float parts.

Fig. 3-40. Pump well and spring.

Step 21. Drop return spring Q into well R (3-40); place accelerator pump plunger assembly on top of spring.

Step 22. Use new gasket; install new needle-valve seats O1 and needle valves O2 (3-39).

Step 23. Place new gasket on cover.

Step 24. Replace both floats N (3-39) with hinge pins M.

Step 25. Adjust float levels. Refer to Fig. 3-41 and sheet instructions in parts kit. *Float level adjustment should be made with care.*

Step 26. Place bowl cover on carburetor bowl; be sure accelerator pump plunger lines up with proper hole in cover. Insert the eight hold-down screws L (3-42) and tighten firmly.

Step 27. Drop return springs J (3-43) into wells K.

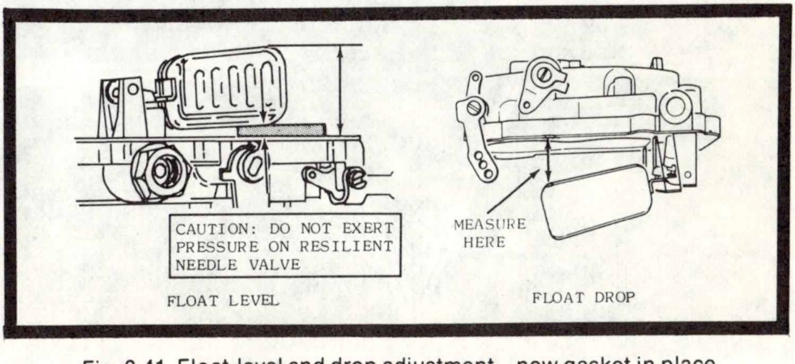

Fig. 3-41. Float-level and drop adjustment—new gasket in place.

Fig. 3-42. Bowl-cover screws.

Step 28. Place metering rods I (3-43) and vacuum-piston assemblies into wells K.

Step 29. Proper alignment of metering rods into jets can be determined by moving assembly up and down with finger against spring pressure, until piston can be easily depressed below level of retainer plate. Hold assembly in depressed position and install retainer plates G with screws F.

Step 30. Install accelerator pump lever E1 (3-44) with screw E. Tighten screws firmly.

Step 31. Install linkage D and retainer clip C (3-44).

Step 32. Use retainer clip A to hook up linkage B (3-44).

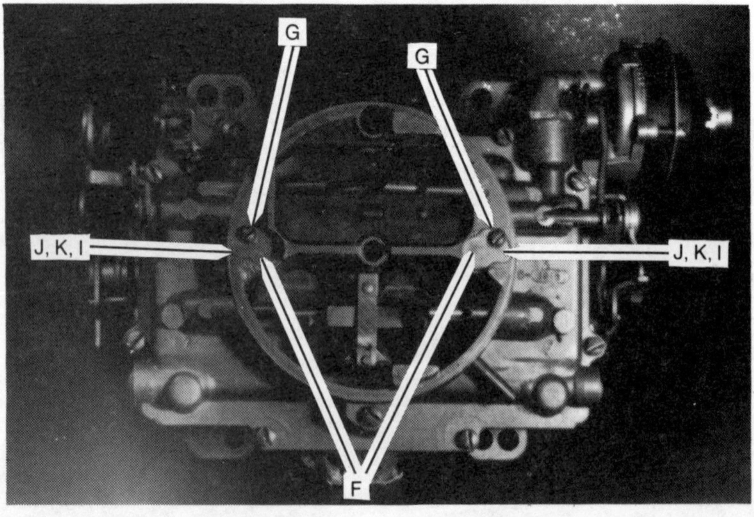

Fig. 3-43. Metering rod and related parts.

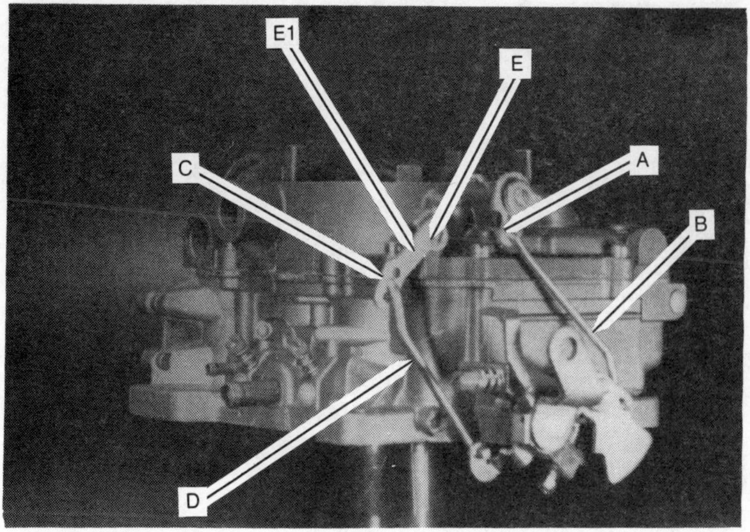

Fig. 3-44. Linkage.

Step 33. Install carburetor on engine. Be sure all lines and linkages are properly connected.

Step 34. Start engine. Allow engine to warm up until choke valve is fully opened.

Step 35. Adjust idle speed with idle speed adjustment screw AA (3-45).

Step 36. Adjust idle mixture with idle mixture screws X (3-45). Turn screws clockwise, one at a time, until engine begins to run rough. Then back screws out counterclockwise until engine runs smoothly. It may be necessary to readjust both screws alternately to obtain smooth engine operation.

Replace air cleaner. Be sure element is clean.

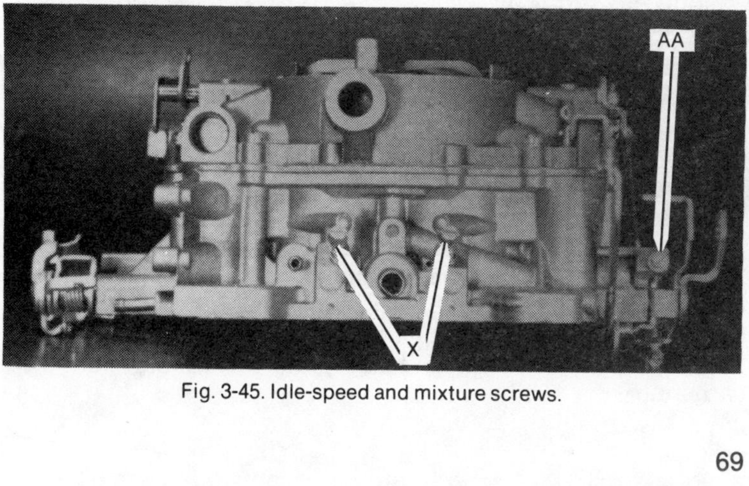

Fig. 3-45. Idle-speed and mixture screws.

Chapter 4

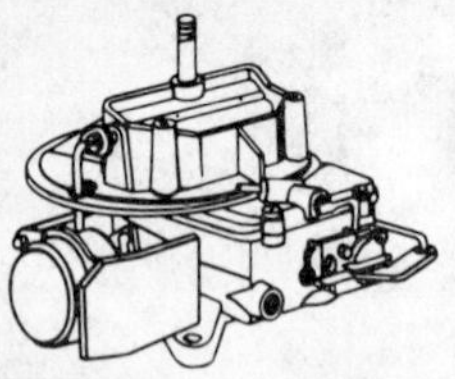

Ford

Ford products have come off the line with Holley, Carter, and, in 1970 and 71, Rochester Quadrajet carburetors. Most production is fitted with Ford's own series that until a few years ago went under the Autolite name. Today these carburetors (and allied products) are marketed as Motorcraft.

Ford carburetors run the gamut from a simple single-barrel to the imposing 4300 four-barrel.

FORD ONE-BARREL CARBURETOR (MODEL F-1)

Please refer to Chapter 1 for a brief description and discussion of carburetor cleaners and the container needed to make up the bath to be used for this task.

Only a few tools are required for tuneup of this one-barrel carburetor, as shown in Table 4-1.

Table 4-1. Tools Required for Ford Carburetor Overhaul.

Screwdriver with a broad blade (approx. $^7/_{16}$ in. or larger).
Screwdriver with narrow blade (approx. $^3/_{16}$ in.).
Needle-nosed pliers.
Adjustable wrench with at least a 1 in. span. or a 1 in. open-end or box-end wrench.
Open-end, box-end, or socket wrench of the proper size to fit the nuts or bolts that hold your carburetor to the manifold.

Note: Most air cleaners can be removed without tools, or with only a pair of pliers.

Preliminary Instructions

Remove the air cleaner.

Remove all dirt and grime from exterior of carburetor while carburetor is still mounted on engine; use kerosene and a paintbrush if necessary, but **do not use gasoline**.

Remove carburetor from engine.

For information on carburetor accessories, refer to Chapter 2. If you're not sure you can remember where and how the lines and linkages connect, make notes and sketches in the margin of this book as you remove the carburetor. Figure 4-1, an exploded view, is included to help you identify the various parts of your carburetor.

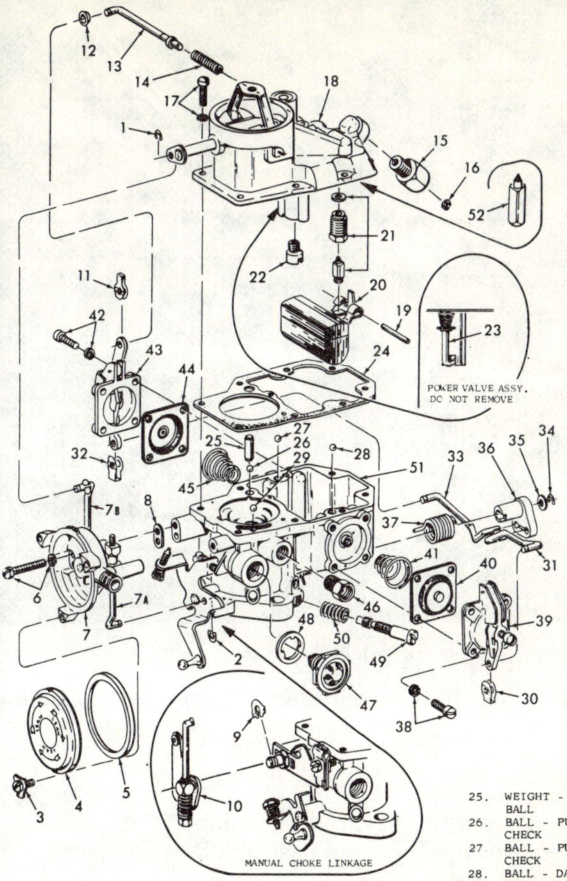

NOMENCLATURE

REF. NO.	
1.	RETAINER - CHOKE ROD
2.	RETAINER - PULL-DOWN ROD
3.	SCREW AND RETAINER-STAT COVER
4.	STAT COVER & SPRING ASSY.
5.	GASKET - STAT COVER
6.	SCREW & LOCKWASHER - CHOKE HOUSING
7.	CHOKE HOUSING ASSY.
7A.	ROD - CHOKE PULL-DOWN
7B.	ROD - CHOKE
8.	GASKET - CHOKE HOUSING
9.	RETAINER - FAST IDLE CAM LEVER & ROD ASSY.
10.	FAST IDLE CAM LEVER & ROD ASSY.
11.	RETAINER - VENT VALVE ROD.
12.	BUSHING - VENT VALVE ROD
13.	ROD - VENT VALVE
14.	SPRING - VENT VALVE ROD
15.	FITTING - FUEL INLET ASSY.
16.	SCREEN - FUEL INLET FITTING
17.	SCREW & LOCKWASHER - BOWL COVER
18.	COVER - BOWL ASSY.
19.	PIN - FLOAT LEVER HINGE
20.	FLOAT & LEVER ASSY.
21.	NEEDLE, SEAT & GASKET ASSY.
22.	JET - MAIN METERING
23.	POWER VALVE ASSY. - DO NOT REMOVE
24.	GASKET - BOWL COVER
25.	WEIGHT - PUMP DISCHARGE BALL
26.	BALL - PUMP DISCHARGE CHECK
27	BALL - PUMP INTAKE CHECK
28.	BALL - DASHPOT CHECK
29.	BALL - DIST. CHECK 223" eng. PARCEL DEL., 262" eng. TRUCK
30.	RETAINER - DASHPOT ROD
31.	ROD - DASHPOT
32.	RETAINER - PUMP ROD
33.	ROD - PUMP
34.	RETAINER - PUMP LEVER
35.	WASHER - PUMP LEVER
36.	LEVER - PUMP
37.	SPRING - OVERTRAVEL
38	SCREW & LOCKWASHER - DASHPOT COVER
39	COVER - DASHPOT & LEVER ASSY.
40	DIAPHRAGM - DASHPOT
41.	SPRING - DASHPOT RETURN
42.	SCREW & LOCKWASHER PUMP COVER
43.	COVER - PUMP & LEVER ASSY.
44.	DIAPHRAGM - PUMP
45.	SPRING - PUMP RETURN
46.	FITTING - DIST. LINE
47.	VALVE - SPARK CONTROL
48.	GASKET - SPARK VALVE
49.	NEEDLE - IDLE ADJ.
50.	SPRING - IDLE ADJ. NEEDLE
51.	FLOAT BOWL & LOWER BODY ASSY.
52.	NEEDLE VALVE (ONLY) **1965** AND UP

Fig. 4-1. Exploded view of Ford one-barrel carburetor, Model F-1.

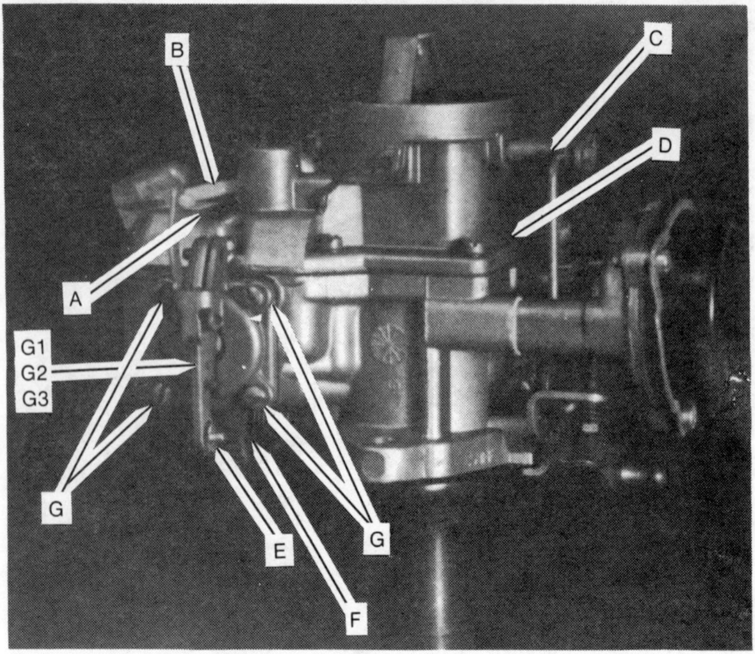

Fig. 4-2. External linkage.

When carburetor is removed from engine, follow tuneup steps listed below.

To Disassemble

Step 1. Remove clip A (4-2); release link B at that point.*
Step 2. Remove clip C (4-2); release link D at that point.
Step 3. Remove clip E (3-2); release kink F at that point.
Step 4. Remove four screws G (4-2); lift off dash-pot cover G1 and remove diaphragm G2 and spring G3 from well G4. Place spring in bath. Discard diaphragm.
Step 5. Remove power valve H (4-3) from well H1—use any wrench with a 1 in. bite. Discard valve.
Step 6. Remove spring clip I (4-3) and release linkage I1 from lever 12.
Step 7. Remove four screws J (4-3); lift off accelerator pump cover K and remove diaphragm K1 and spring K2 from well K3. Place spring in bath. Discard diaphragm.
Step 8. Remove three screws L (4-3), retainer ring M, and choke housing N. Do *not* place in bath.
Step 9. Remove eight bowl-cover screws O (4-3); lift cover off carburetor bowl. Discard gasket.

---

*On some models, this linkage is released by removing a screw at the lower end where it attaches to the fast-idle cam.

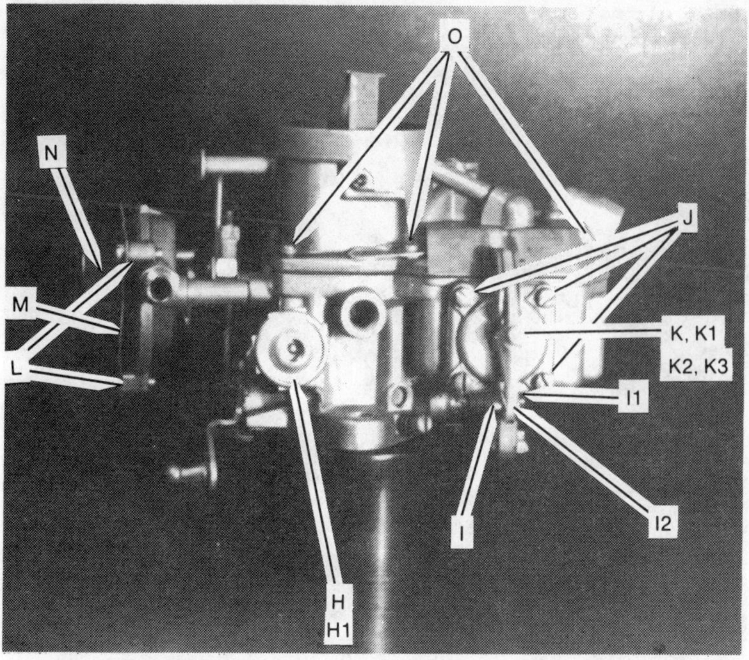

Fig. 4-3. Power valve, accelerator pump, and choke.

Step 10. Remove float-hinge pin P (4-4) and lift float Q. Do *not* place in bath.

Step 11. Remove needle valve R (4-4) from seat R1. Discard valve.

Step 12. Place bowl cover in bath.

Step 13. Cup your hand over wells S (4-5), and S1, and T, invert bowl and allow balls S2 and S3 and weight T1 to drop into hand. Place weight in bath. Discard the balls.

Step 14. Remove idle-mixture adjustment screw U (4-5) and spring V; place both in bath.

Step 15. Place carburetor bowl in bath.

Note: Disassembly is complete. Allow all parts placed in bath to be properly cleaned in accordance with instructions on carburetor-cleaner container before beginning assembly. Then remove all parts from bath, rinse with water, and allow to dry naturally; do not use compressed air.

To Reassemble

Step 16. Install idle-mixture adjustment screw U (4-5) into spring V; run screw into well V1 until it seats lightly, then back it out 1½ turns.

Step 17. Drop new steel ball T2 (4-5) into well T; place weight T1 on top of ball.

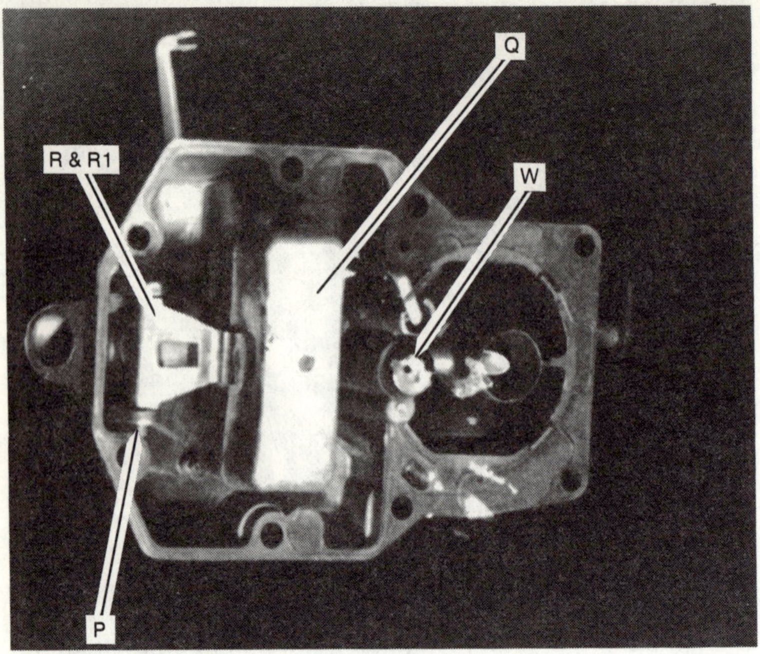

Fig. 4-4. Float and related parts.

Step 18. Drop new steel balls S2 and S3 (4-5) into wells S and S1.

Step 19. Drop new needle valve R (4-6) point first into seat R1.

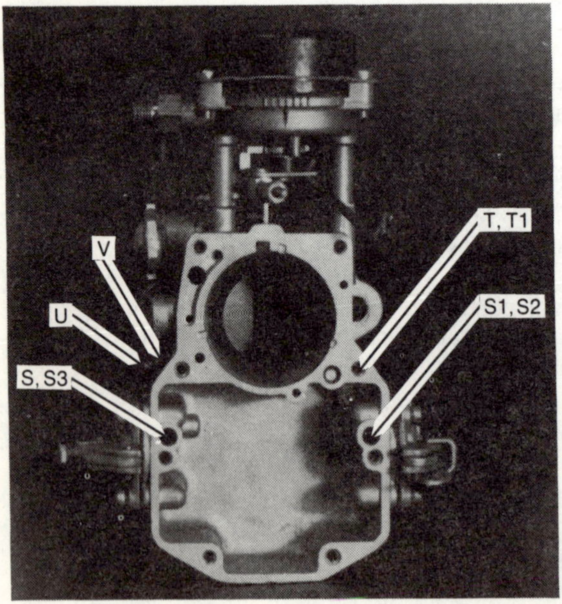

Fig. 4-5. Wells and check valves.

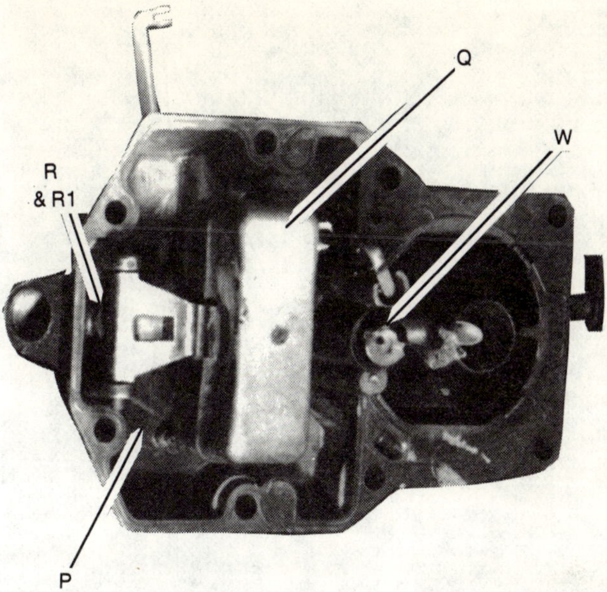

Fig. 4-6. Float parts and main jet.

Step 20. Reinstall float Q (4-6) into bowl cover with hinge pin P.

Step 21. Adjust float level. Refer to instructions included in your parts kit.

Step 22. Use broad-blade screwdriver to tighten jet W (4-6).

Step 23. Place new gasket into proper position on bowl.

Step 24. Lower bowl cover into position on bowl and attach with eight screws O (4-7). Tighten screws firmly.

Step 25. Install new diaphragm K1 (4-7) and tapered coil spring K2 into accelerator pump well K3.

Note: Install spring point-first then place diaphragm over spring with "button" out.

Step 26. Install accelerator pump cover K (4-7) with four screws J. Be sure screws are aligned with diaphragm. Tighten screws.

Step 27. Attach linkage I1 (4-7) to lever I2 with clip I.

Step 28. Use new gasket. Install new power valve H (4-7) into well H1—tighten it, but keep in mind size of wrench being used and kind of metal carburetor is made of.

Step 29. Place new diaphragm G2 (4-8), "button" side out, over big end of tapered spring G3, then insert both into dash-pot well G4, spring first.

Step 30. Place dash-pot cover G1 (4-8) into position with lever hinge toward top of carburetor; install with four screws

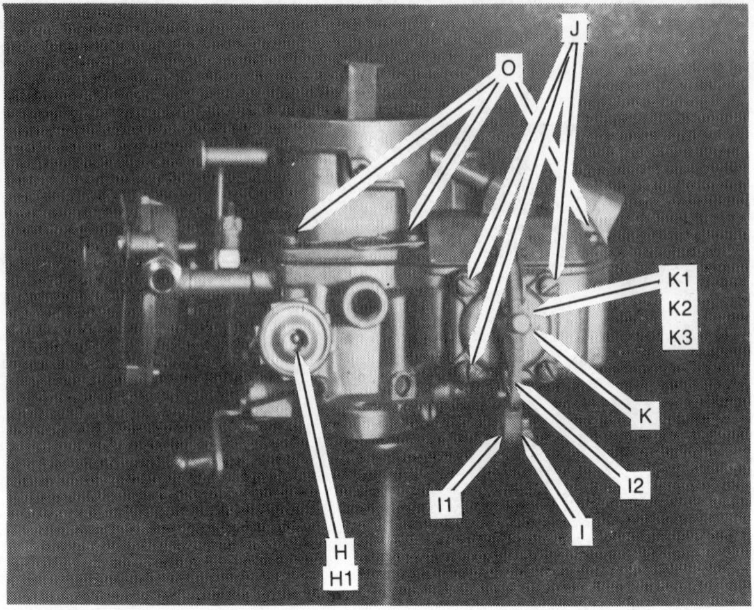

Fig. 4-7. External parts including power valve and accelerator.

G. Be sure screws are aligned with diaphragm. Tighten screws.

Step 31. Attach linkage F (4-8) at point F1 with clip E.
Step 32. Attach linkage D (4-8) at point D1 with clip C.

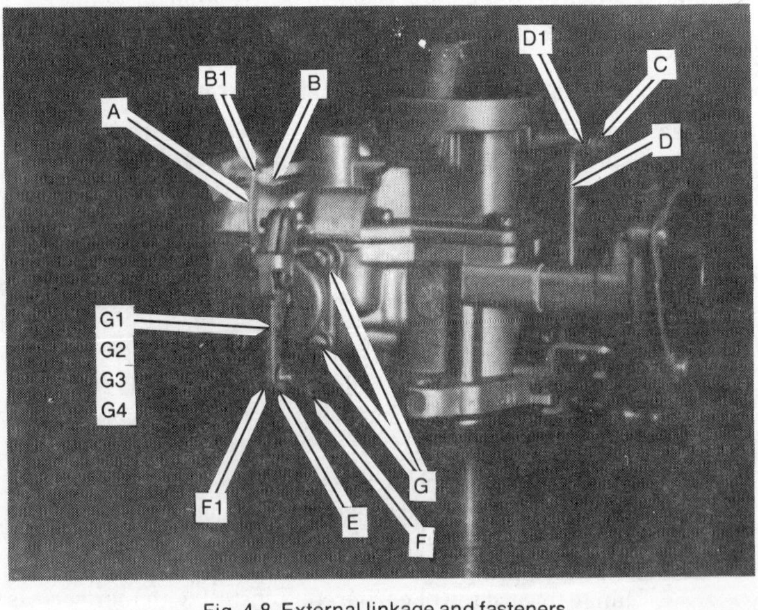

Fig. 4-8. External linkage and fasteners.

Step 33. Attach linkage B (4-8) at point B1 with clip A.

Step 34. Install carburetor on engine. Be sure fuel line is tightly connected and all linkages are properly attached.

Step 35. Start engine. Allow it to warm up until choke butterfly in throat of carburetor is fully open.

Step 36. Adjust idle to desired speed with screw X (4-9).

Step 37. Adjust idle mixture by turning screw U (4-9) clockwise until engine begins to falter, then counterclockwise until smooth idle is achieved.

Mount air cleaner to carburetor. Be sure element is clean.

FORD TWO-BARREL CARBURETOR (MODEL 2100)

Please refer to Chapter 1 for a brief description and discussion of carburetor cleaners and the solvent container needed to make the bath to be used for this task.

Only a few tools are required for tuneup of this two-barrel carburetor; they are listed in Table 4-1.

Note: Most air cleaners can be removed without tools or with only a pair of pliers.

Preliminary Instructions

Remove the air cleaner.

Remove all dirt and grime from exterior of carburetor while carburetor is still mounted on engine; use kerosene and a paintbrush if necessary, but **do not use gasoline**.

Remove carburetor from engine.

For information on carburetor accessories, refer to Chapter 2. If you are not sure you can remember where and how the lines

Fig. 4-9. Adjustment screws.

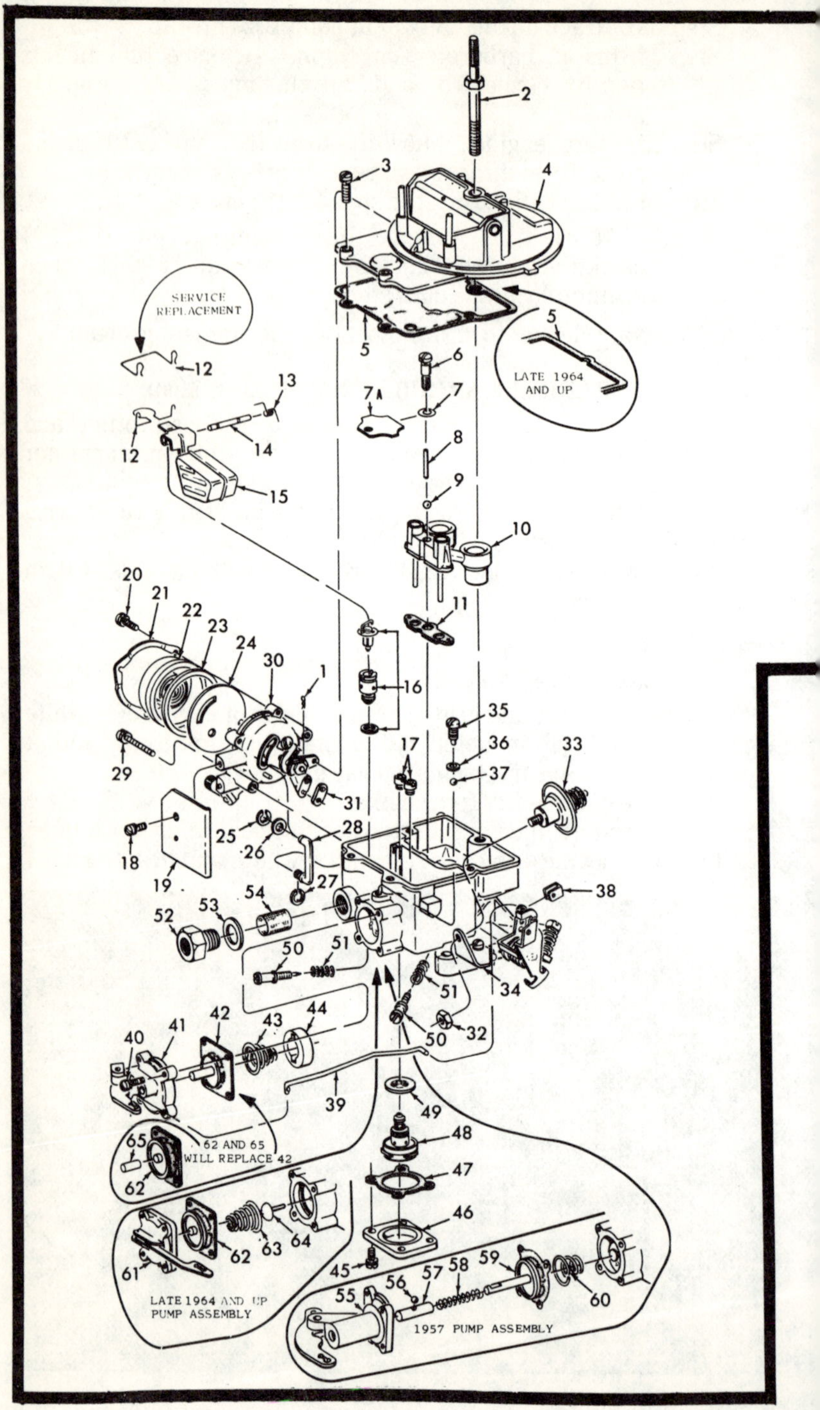
SERVICE REPLACEMENT
LATE 1964 AND UP
62 AND 65 WILL REPLACE 42
LATE 1964 AND UP PUMP ASSEMBLY
1957 PUMP ASSEMBLY

NOMENCLATURE

REF. NO.		REF. NO.	
1.	RETAINER-CHOKE ROD LOWER	33.	DASHPOT
2.	STUD-AIR CLEANER	34.	BRACKET-DASHPOT
3.	SCREW-BOWL COVER	35.	SCREW-INLET CHECK BALL RETAINER
4.	BOWL COVER ASSY.	36.	GASKET-INLET CHECK BALL RETAINER SCREW
5.	GASKET-BOWL COVER	37.	BALL-PUMP INLET CHECK
6.	SCREW-PUMP DISCHARGE NOZZLE	38.	RETAINER-PUMP ROD
7.	GASKET-PUMP DISCHARGE NOZZLE SCREW	39.	PUMP ROD
7A	PLATE (AIR DISTRIBUTION)-PUMP DISCHARGE NOZZLE. LATE 1964 & 1965 PARTIAL PRODUCTION	40.	SCREW & LOCKWASHER-PUMP COVER
8.	WEIGHT-DISCHARGE CHECK BALL	41.	PUMP COVER & LEVER ASSY.
9.	BALL-DISCHARGE CHECK	42.	PUMP DIAPHRAGM ASSY.
10.	VENTURI CLUSTER ASSY.	43.	SPRING-PUMP DIAPHRAGM RETURN
11.	GASKET-VENTURI CLUSTER ASSY.	44.	CAVITY FILLER-PUMP
12.	RETAINER-FLOAT PIN	45.	SCREW & LOCKWASHER-ECONOMIZER VALVE COVER
13.	SPRING-FLOAT DAMPNER	46.	COVER-ECONOMIZER VALVE
14.	PIN-FLOAT HINGE	47.	GASKET-ECONOMIZER VALVE COVER
15.	FLOAT & LEVER ASSY.	48.	VALVE-ECONOMIZER
16.	NEEDLE, SEAT & GASKET ASSY.	49.	GASKET-ECONOMIZER VALVE
17.	JETS-MAIN METERING	50.	NEEDLES-IDLE ADJUSTING
18.	SCREW & LOCKWASHER-AIR SHIELD	51.	SPRINGS-IDLE ADJUSTING NEEDLE
19.	AIR-SHIELD	52.	FITTING-FUEL INLET
20.	SCREW & LOCKWASHER-STAT CLAMP	53.	GASKET-FUEL INLET FITTING
21.	CLAMP-STAT COVER	54.	SCREEN-FUEL INLET FILTER
22.	STAT COVER & SPRING ASSY.	55.	PUMP COVER & LEVER ASSY.-1957
23.	GASKET-STAT COVER	56.	BALL-PUMP PUSH ROD SLEEVE-1957
24.	PLATE-CHOKE BAFFLE	57.	SLEEVE-PUMP PUSH ROD-1957
25.	RETAINER-FAST IDLE ROD UPPER	58.	SPRING-PUMP PUSH ROD-1957
26.	WASHER-FAST IDLE ROD UPPER	59.	PUMP DIAPHRAGM ASSY.-1957
27.	RETAINER-FAST IDLE ROD LOWER	60.	SPRING-PUMP DIAPHRAGM RETURN-1957
28.	FAST IDLE ROD	61.	PUMP COVER & LEVER ASSY.
29.	SCREW & LOCKWASHER-CHOKE HOUSING	62.	PUMP DIAPHRAGM ASSY.
30.	CHOKE HOUSING ASSY.	63.	SPRING-PUMP DIAPHRAGM RETURN
31.	GASKET-CHOKE HOUSING ASSY.	64.	VALVE-PUMP INLET CHECK
32.	LOCKNUT-DASHPOT	65.	PUSH ROD-PUMP DIAPHRAGM

Fig. 4-10. Exploded view of Ford two-barrel carburetor, Model 2100.

and linkages connect, make notes and sketches in the margin of this book as you remove the carburetor. Figure 4-10, an exploded view, is included to aid you in the identification of the several parts of your carburetor.

When carburetor is removed from engine, follow tuneup steps listed below.

To Disassemble

Step 1. Remove retainer A (4-11) and release linkage B.

Step 2. Remove the four screws C (4-12); lift cover off carburetor bowl.

Step 3. Remove and discard bowl-cover gasket.

Step 4. Release spring retainer E (4-13); lift float F from bowl.

Step 5. Remove needle valve G and seat H (4-13) from bowl.

Step 6. Remove single screw I (4-13) and lift discharge-nozzle assembly J from throat of carburetor.

Step 7. Beneath screw I (which you just removed) is a pin weight K holding a check ball L. Remove both pin and ball. Discard ball.

Step 8. Remove linkage M (4-13).

Step 9. Remove four screws N holding accelerator pump cover O and diaphragm assembly P (4-12). Remove

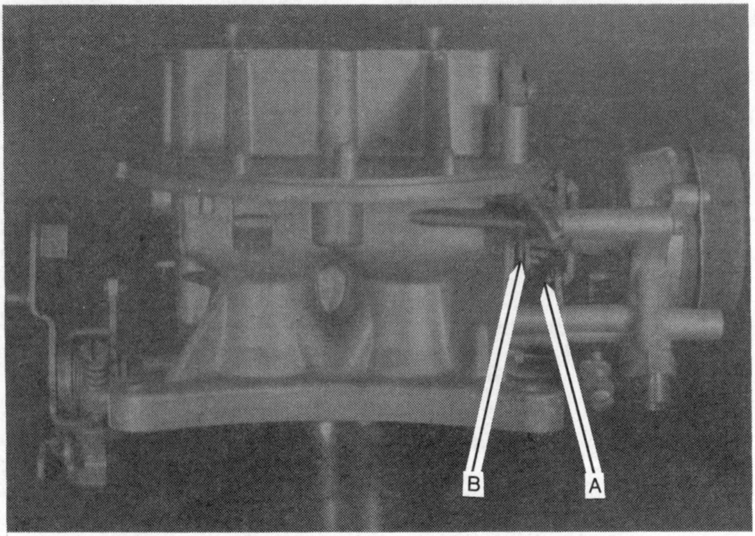

Fig. 4-11. Linkage.

cover, diaphragm assembly, and spring Q. Discard the diaphragm. Place cover and spring in bath.

Step 10. Remove four screws R (4-14). Lift off housing S that covers power valve T. Discard gasket.

Step,11. Remove power valve T (4-14) with pliers or 1 in. box- or open-end wrench. Discard valve.

Fig. 4-12. External fasteners.

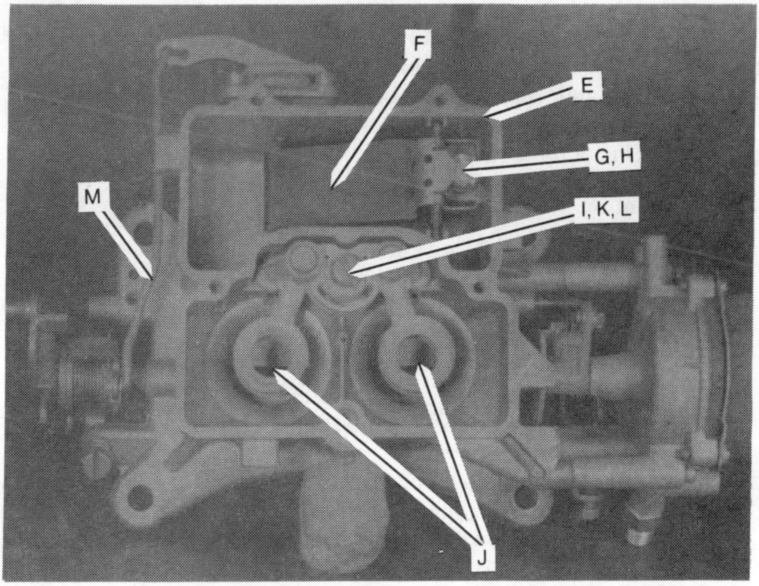

Fig. 4-13. Float assembly and discharge nozzle.

Step 12. Remove the two idle-mixture adjustment screws U (4-14) and springs U1.

Step 13. Remove three screws V and retainer ring V1 that holds black choke housing W (4-14). Remove housing W from carburetor.

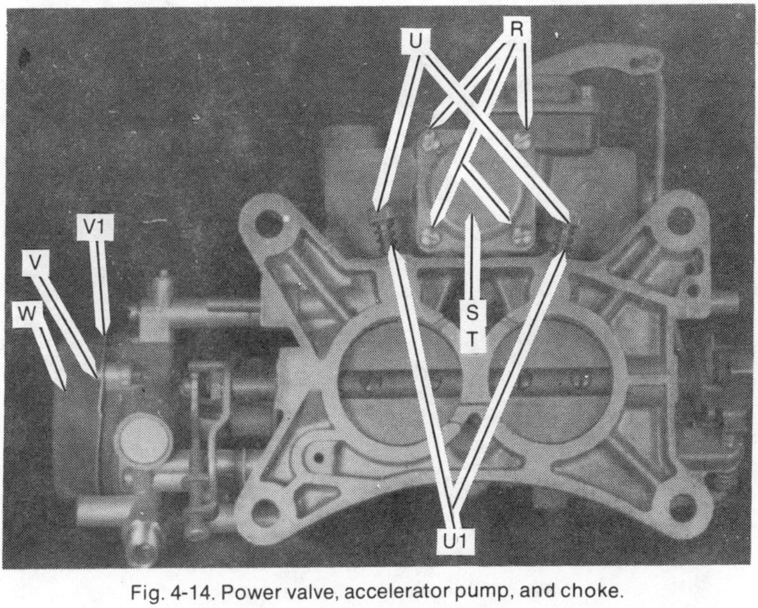

Fig. 4-14. Power valve, accelerator pump, and choke.

Note: Place all metal parts in bath. Allow all parts placed in bath to be properly cleaned in accordance with directions on carburetor-cleaner container before beginning reassembly. Remove all parts from cleaner, rinse with water, and allow to dry naturally. Do not use compressed air.

To Reassemble

Step 14. Install new power valve T (4-14).

Step 15. Place new gasket on power-valve cover S and place cover over power valve T (4-14). Tighten the four screws R firmly.

Step 16. Install the two idle-mixture adjustment screws U (4-14) and springs U1 into wells U2. Run screws all the way in until you feel needle seat snugly—not tight—then back out a full turn.

Step 17. Replace accelerator pump assembly (4-15). Place small end of cone-shaped spring Q into housing of carburetor at Q1. Place new diaphragm assembly P into cover O, then attach cover with four screws N. Tighten screws firmly.

Step 18. Replace linkage M (4-16).

Step 19. Drop new check ball L into well L1 (4-16). Drop pin weight K on top of check ball.

Step 20. Replace discharge-nozzle assembly J (4-16). Place new gasket under head of single screw I. Tighten screw.

Step 21. Install new needle valve G and seat H (4-16). Tighten seat firmly.

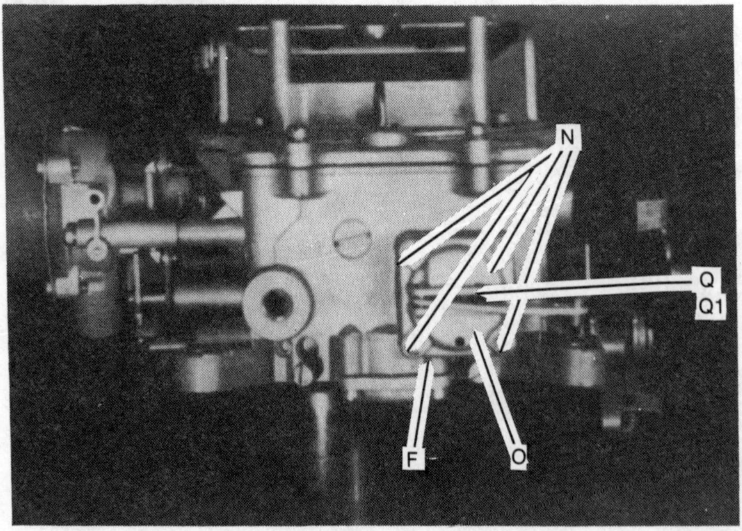

Fig. 4-15. Accelerator pump assembly.

Fig. 4-16. Float, needle and seat, and discharge-nozzle cluster.

Step 22. Install float F (4-16). Be sure to place long end of small coil spring E down into bowl of carburetor; place crooked end underneath float. This spring assists float to stay afloat on gasoline. Hook spring in back of needle-valve seat.

Step 23. Make float-level adjustment in accordance with instructions included with parts kit.

Step 24. Use new gasket; replace bowl cover D (4-17). Tighten the four screws C firmly.

Step 25. Attach choke linkage B (4-18) with clip A.

Step 26. Attach black choke housing W with three screws V and retainer V1. Before tightening screws, rotate housing until choke butterfly closes and single mark on black housing lines up with center line on the metal housing.

Step 27. Install carburetor on engine. Be sure all linkage and lines are properly attached.

Step 28. Start engine. Allow it to warm up until choke valve is fully open.

Step 29. Adjust idle speed by rotating screw X (4-18) until desired idle speed is achieved.

Step 30. Adjust idle mixture with two screws U (4-17). Turn each screw clockwise until engine begins to falter, then turn counterclockwise until engine runs smoothly. Adjust screws alternately until smoothest idle is achieved.

Replace air cleaner on carburetor. Be sure element is clean.

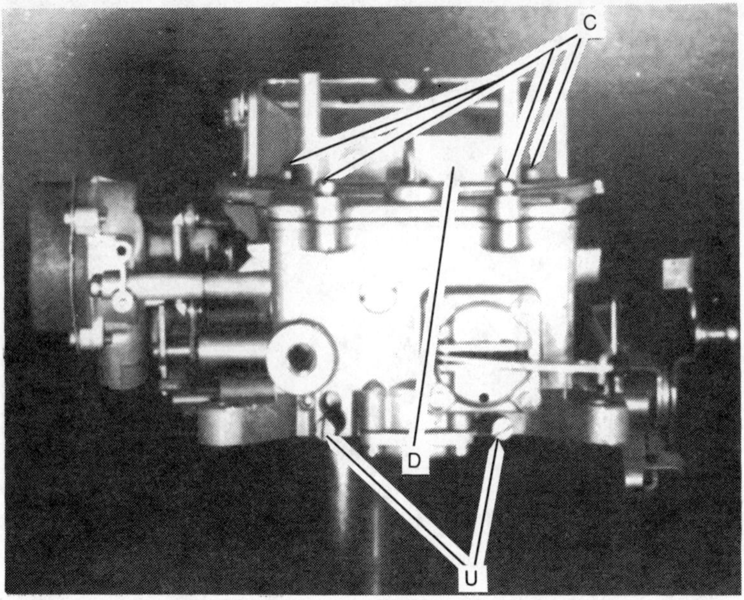

Fig. 4-17. Bowl-cover and mixture-adjustment screws.

FORD FOUR-BARREL CARBURETOR (MODEL 4100)

Please refer to Chapter 1 for a brief description and discussion of carburetor cleaners and the container needed to make up the bath to be used for this task.

Only a few tools are needed for tuneup of this four-barrel carburetor, as shown in Table 4-1.

Note: Most air cleaners can be removed without tools or with a pair of pliers.

Fig. 4-18. Linkage, idle-speed screw, and choke housing.

Preliminary Instructions

Remove air cleaner.

Remove all dirt and grime from exterior of carburetor while carburetor is still mounted on engine; use kerosene and a paintbrush if necessary, but **do not use gasoline**.

Remove carburetor from engine.

For information on carburetor accessories, refer to Chapter 2. If you are not sure you can remember where and how the lines and linkages connect, make notes and sketches in the margin of this book as you remove the carburetor. Figure 4-19, an exploded view, is included to help you identify the various parts of your carburetor.

When carburetor is removed from engine, follow tuneup steps listed below.

To Disassemble

Step 1. Remove clip A (4-20) and detach linkage B.

Step 2. Remove eight screws C (4-20); lift bowl cover off bowl. Discard gasket.

Step 3. Release retainer springs D (4-21). Lift both floats E and small coil springs DD from bowl.

Step 4. Remove the two needle-and-seat assemblies F (4-21). Discard these assemblies.

Step 5. Position carburetor before you as shown in Fig. 4-21. The right-hand half is the *primary* portion. The left-hand half is the *secondary* portion. In normal driving the primary portion does most of the work. The secondary portion operates only under certain load and speed conditions.

Step 6. Remove screw G (4-21). Lift primary discharge assembly H from throat of carburetor. Remove pin weight I and check ball J from well K directly under screw G.

Step 7. Remove screw L (4-21). Lift secondary discharge assembly M from throat of carburetor.

Note: The secondary discharge assembly is held in place by a solid screw L. The primary assembly is held in place by a hollow screw G. Also, the primary assembly has discharge holes in casting at N (4-21). Keep this information in mind when assembling the carburetor.

Step 8. Remove linkage O (4-22).

Step 9. Remove four screws P (4-22). Lift off accelerator pump housing P1, diaphragm P2, and cone-shaped spring P3.

Step 10. Remove linkage Q (4-22).

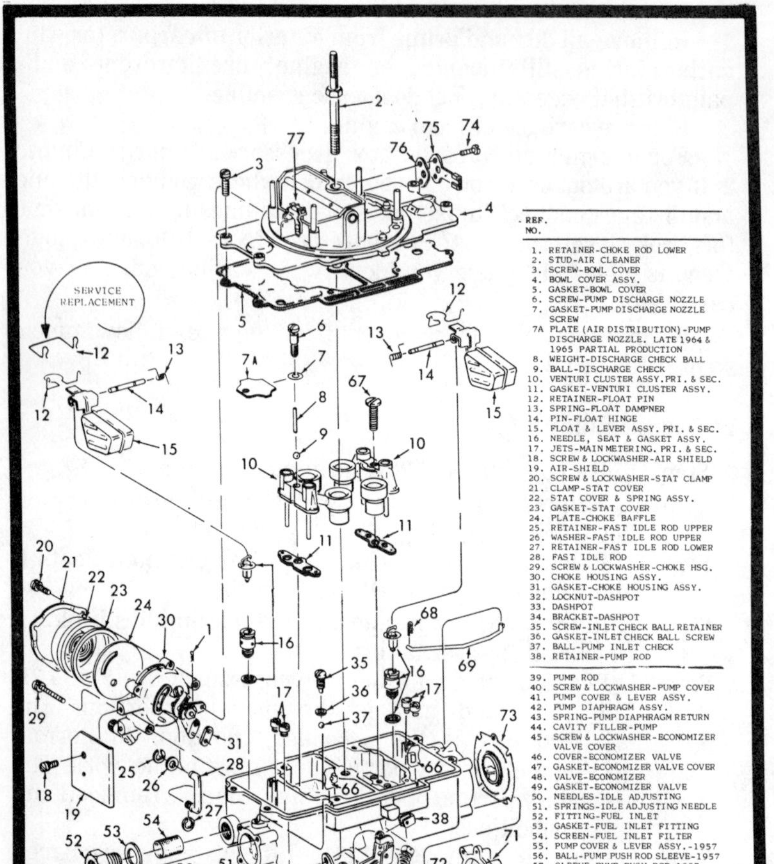

Fig. 4-19. Exploded view of Ford four-barrel carburetor, Model 4100.

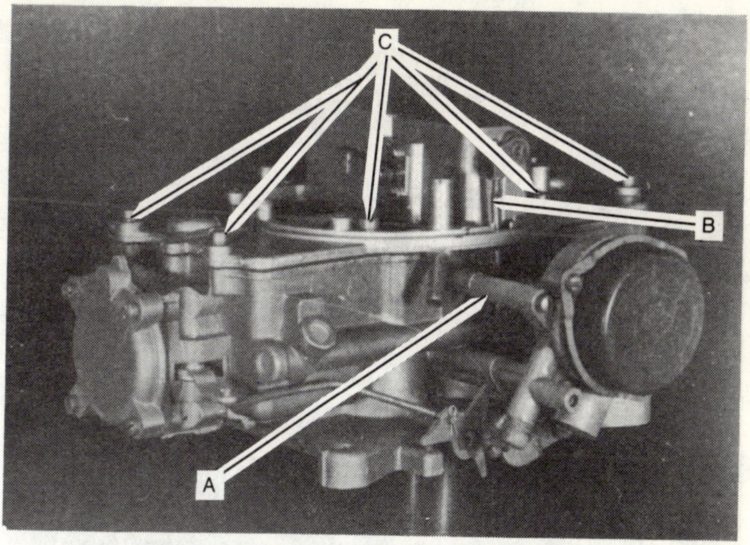

Fig. 4-20. Bowl-cover screws, choke, and linkage.

Step 11. Remove four screws R (4-22). Lift off housing R1 and diaphragm R2 of secondary vacuum assembly.

Step 12. Remove four screws S (4-22). Remove housing S1 that covers power valve T.

Step 13. Remove power valve T (4-22).

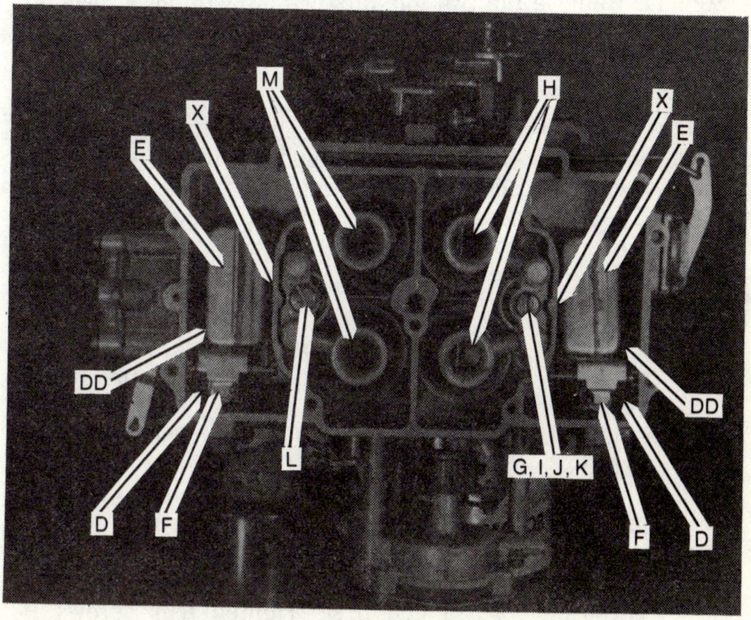

Fig. 4-21. Bowl assemblies and discharge nozzles.

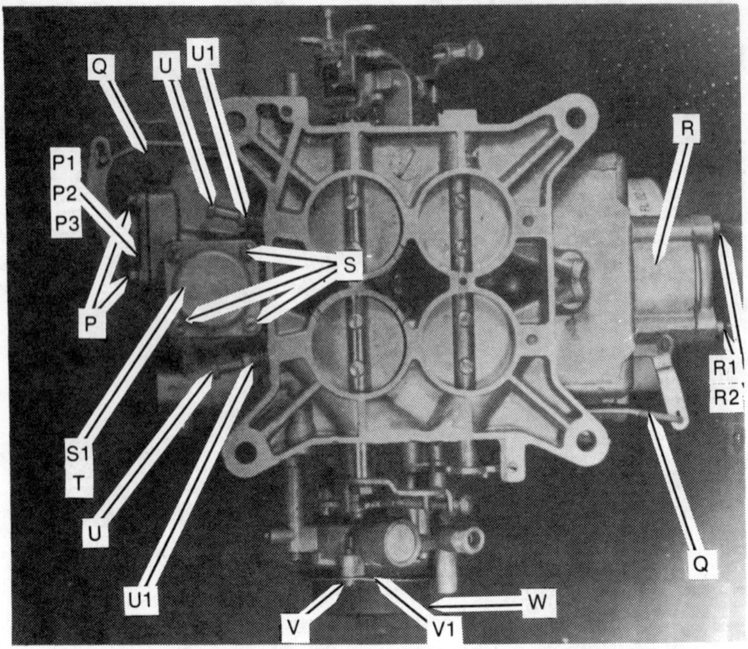

Fig. 4-22. Power valve, vacuum assembly, accelerator pump, and miscellaneous parts.

Step 14. Remove the two idle-mixture adjustment screws U and springs U1 (4-22).

Step 15. Remove the three screws V and retainer V1 that hold choke housing W (4-22) in place.

Note: Place all metal parts in bath. Allow all parts placed in bath to be properly cleaned in accordance with directions on carburetor-cleaner container before beginning assembly. Then, remove all parts from cleaner, rinse with water, and allow to dry naturally. Do not use compressed air.

To Reassemble

Step 16. Replace two idle-mixture adjustment screws and springs U and U1 (4-22). Turn both screws in firmly but only finger-tight.

Step 17. Install new power valve T (4-22) and gasket. Tighten firmly.

Step 18. Replace cover S1 (4-22) over power valve T—use new gasket. Tighten the four cover screws S firmly.

Step 19. Place new accelerator pump diaphragm P2 (4-22) into housing P1. Place cone-shaped spring P3 into carburetor, small end first. Tighten four screws P firmly.

Step 20. Replace linkage O (4-22).

Step 21. Install secondary-diaphragm assembly R2 (4-22) and cover R1. Tighten the four screws R firmly.

Step 22. Replace linkage O (4-22).

Step 23. Replace secondary-discharge nozzle assembly M (4-23). Use the solid screw L and a new gasket. Tighten screw firmly.

Step 24. Place primary discharge nozzle assembly H (4-23) into position. Use new gasket. Drop check ball J into screw hole K; drop pin weight I on top of check ball; use new gasket under hollow screw G. Tighten firmly.

Step 25. Tighten jets X (4-23) in bottom of both fuel bowls.

Step 26. Install new needle-and-seat assemblies F (4-23). Tighten firmly.

Step 27. Install both floats E as viewed in 4-23. The small coil springs DD are positioned with hooks underneath floats, while the straight ends point downward into bowl. These springs lift up on floats to aid them in floating. Hook the U-shaped springs D around needle-valve seats to hold float assemblies in position.

Step 28. Set float level according to directions included with parts kit. *Be accurate*.

Step 29. Replace bowl cover. Use new gasket. Tighten all eight cover screws C (4-24) firmly.

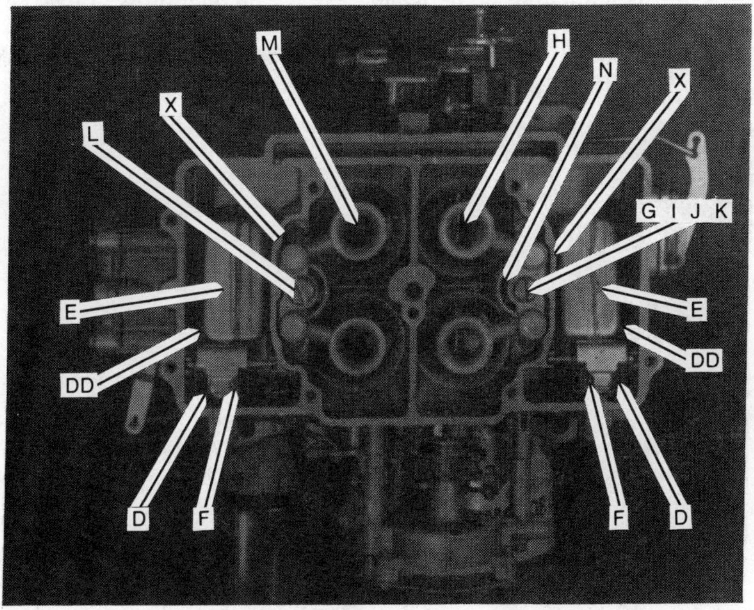

Fig. 4-23. Float, nozzle clusters, and related hardware.

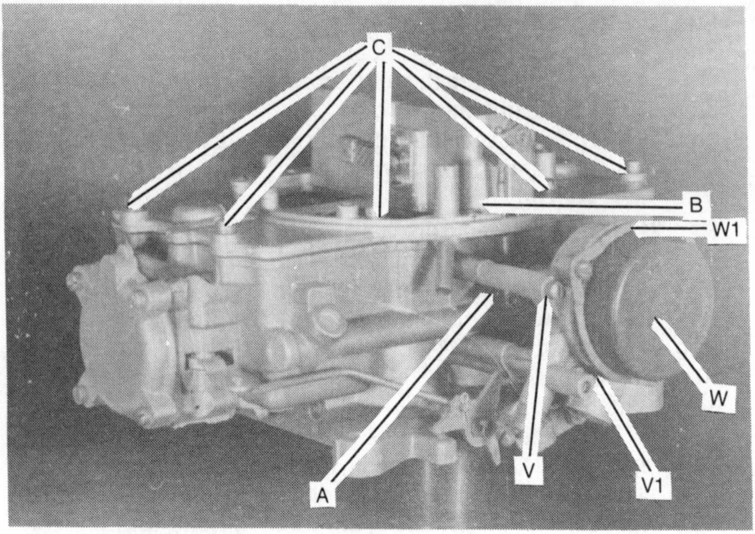

Fig. 4-24. Cover assembly.

Step 30. Attach linkage B (4-24), and replace clip A.

Step 31. Replace choke housing W (4-24). Fasten in place with retainer V1 and three screws V. Before tightening screws, rotate housing counterclockwise until choke valve in carburetor throat closes, and black mark on housing lines up with center mark on casting W1 (4-24). Tighten screws firmly.

Step 32. Install carburetor on engine. Be sure all lines and linkages are properly connected.

Fig. 4-25. Idle-speed adjustment screw.

Step 33. Start engine. Allow engine to warm up until choke valve opens fully.

Step 34. Adjust idle speed with screw Y (4-25).

Step 35. Adjust idle mixture with screws U. Turn each screw clockwise until engine begins to falter, then back out until engine runs smoothly. Alternate back and forth between screws, adjusting until smoothest idle is obtained.

Replace the air cleaner on the carburetor. Be sure element is clean.

FORD FOUR-BARREL CARBURETOR (MODEL 4300)

Please refer to Chapter 1 for a brief description and discussion of carburetor cleaners and the container needed to make up the bath to be used for this task. Refer to Table 4-1 for tools needed. In addition to these listed, you'll need one medium Phillips.

Preliminary Instructions

Remove air cleaner.

Remove all dirt and grime from exterior of carburetor while carburetor is still mounted on engine; use kerosene and a paintbrush if necessary, but **do not use gasoline.**

Remove carburetor from engine.

For information on carburetor accessories, refer to Chapter 2. If you are not sure you can remember where and how the lines

Fig. 4-26. Idle-mixture screw—one of two shown.

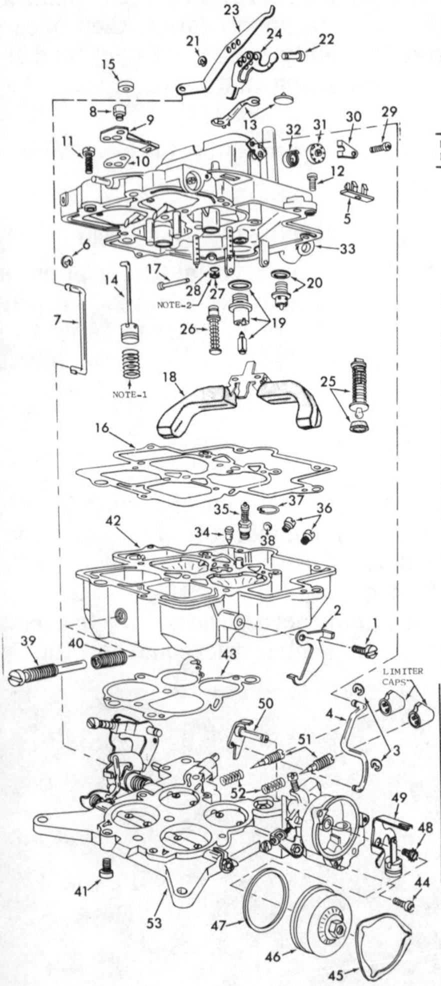

NOMENCLATURE

REF. NO.	
1.	SCREW - SECONDARY LOCKOUT LEVER
2.	LEVER - SECONDARY LOCKOUT
3.	RETAINERS (2) CHOKE ROD
4.	ROD - CHOKE
5.	SEAL - CHOKE ROD DUST
6.	RETAINER - PUMP ROD
7.	ROD - PUMP
8.	SCREW (2) HOT IDLE COMPENSATOR VALVE
9.	VALVE - HOT IDLE COMPENSATOR
10.	GASKET - HOT IDLE COMPENSATOR VALVE
11.	SCREW - (1) BOWL COVER
12.	SCREW - (10) BOWL COVER
13.	VALVE - IDLE VENT
14.	PISTON & ROD ASSY. - AIR VALVE
15.	WASHER- AIR VALVE ROD
16.	GASKET - BOWL COVER
17.	PIN - FLOAT HINGE
18.	FLOAT & LEVER ASSEMBLY
19.	NEEDLE,SEAT & GASKET ASSY. PRI.
20.	NEEDLE,SEAT & GASKET ASSY. SEC.
21.	RETAINER - PUMP LEVER PIN
22.	PIN - PUMP LEVER
23.	LEVER - PUMP
24.	LEVER - IDLE VENT VALVE
25.	PUMP ASSEMBLY
26.	POWER PISTON ASSEMBLY
27.	RETAINER - PUMP AIR BLEED VALVE
28.	VALVE - PUMP AIR BLEED
29.	SCREW - AIR VALVE SPRING HOUSING CLAMP
30.	CLAMP - AIR VALVE SPRING HOUSING
31.	HOUSING - AIR VALVE SPRING
32.	SPRING - AIR VALVE 1966-68
33.	BOWL COVER ASSEMBLY
34.	NEEDLE - PUMP DISCHARGE
35.	VALVE - POWER
36.	JETS (2) MAIN
37.	RETAINER - PUMP INTAKE BALL
38.	BALL - PUMP INTAKE
39.	IDLE AIR BYPASS SCREW
40.	SPRING - IDLE AIR BYPASS SCREW
41.	SCREW (6) - THROTTLE BODY
42.	BOWL ASSEMBLY - FLOAT
43.	GASKET - THROTTLE BODY
44.	SCREW (3) - STAT RETAINER
45.	RETAINER - STAT
46.	STAT COVER & SPRING ASSY.
47.	GASKET - STAT COVER
48.	SCREW & LOCKWASHER - CHOKE PISTON & LINK
49.	PISTON & LINK - CHOKE
50.	SHAFT & LEVER - CHOKE HOUSING
51.	NEEDLES - IDLE ADJUSTING
52.	SPRINGS - IDLE ADJ. NEEDLES
53.	THROTTLE BODY ASSEMBLY

NOTE 1: 1969 & LATER-SECONDARY AIR VALVE SPRING. NO ADJUSTMENT REQUIRED.
NOTE 2: REF. NO. 27 & 28 NOT REQUIRED AFTER 1971.

Fig. 4-27. Exploded view of Ford four-barrel carburetor, Model 4300.

and linkages connect, make notes and sketches in the margins of this book as you remove the carburetor. Figure 4-27, an exploded view, is included to help you identify the various parts of your carburetor.

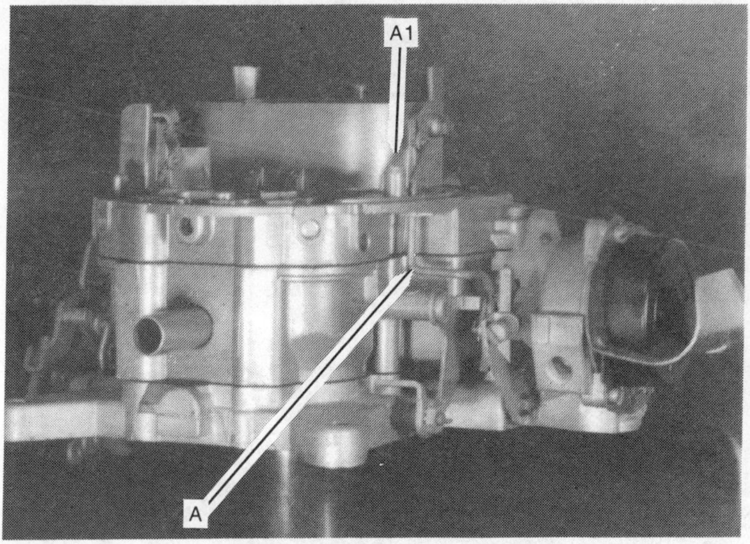

Fig. 4-28. Choke linkage and retaining clip.

When carburetor is removed from engine, follow tuneup steps listed below.

To Disassemble

Step 1. Place carburetor before you as shown in Fig. 4-28.

Step 2. Remove pin clip A1 (4-28); disconnect and remove choke linkage A.

Step 3. Disconnect linkage B at coil spring B1 (4-29) at B2. Place spring in bath.

Step 4. Remove the eleven bowl-cover screws C (4-29), and lift cover off bowl. Discard gasket.

Step 5. From bowl cover, extract lever D5 (4-29), remove air-valve piston D1 (4-30), spring D2, and rod D3 (4-30).

Step 6. Remove float E (4-30) by removing pin E1 at point E2. Avoid applying excessive pressure on float.

Step 7. Remove pin F1 (4-31) from pump-activating arm F2 (on top of bowl cover). Now remove accelerator pump assembly F (4-30). It is not absolutely essential that you remove this assembly, since only the rubber cup F3 on the end of the plunger has to be replaced. However, the plastic retainer F4 at the top of spring F5 may be damaged by bath, so disassembly is recommended.

Step 8. Remove needle valve G (4-30) from well G2; then, with broad-blade screwdriver, remove needle-valve seat G1. Discard both needle and seat.

Step 9. Place bowl cover in bath.

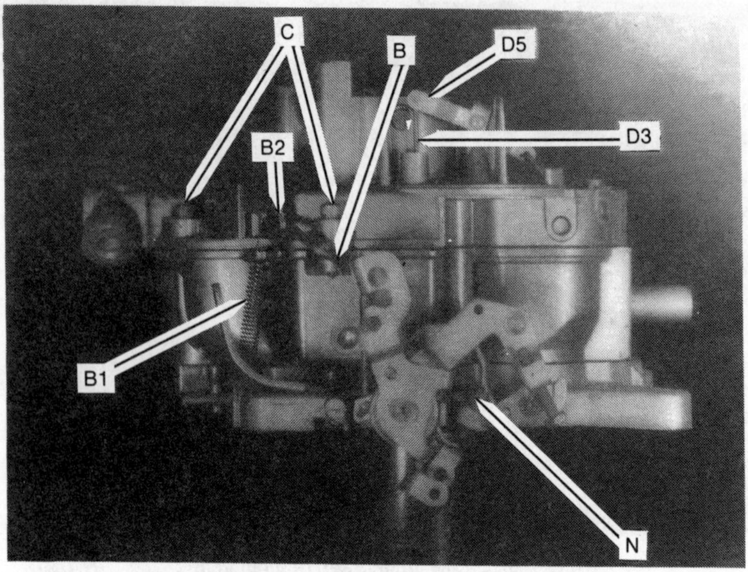

Fig. 4-29. External linkage and fasteners.

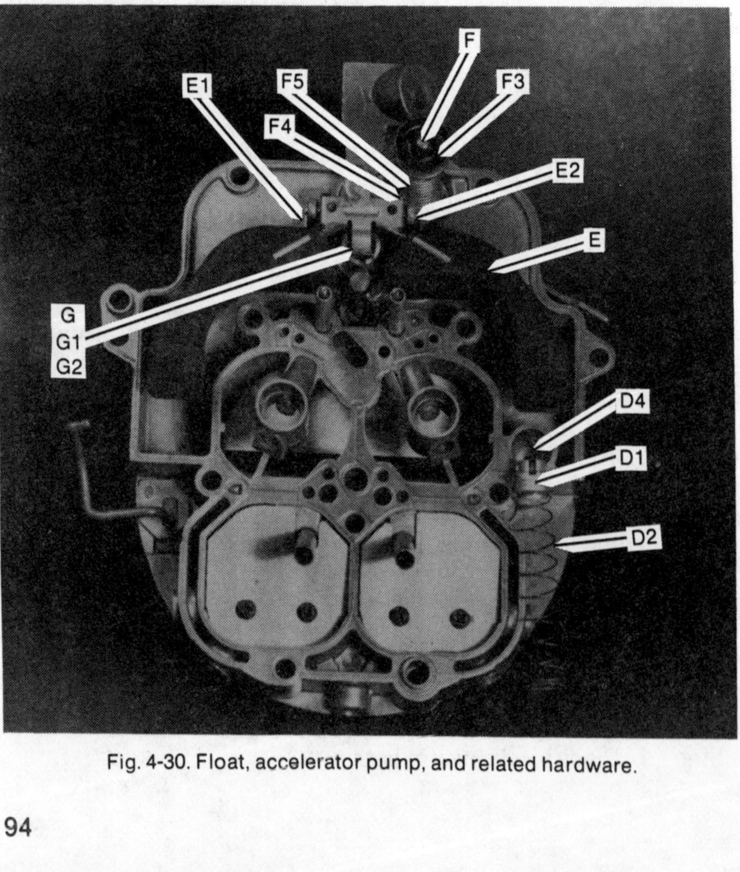

Fig. 4-30. Float, accelerator pump, and related hardware.

Fig. 4-31. Accelerator pump externals.

Step 10. From carburetor bowl, use ⅜ in. deep socket to remove power valve H (4-32).

Step 11. Remove accelerator pump discharge needle valve I (4-32) from well I1—hold hand over well, turn carburetor over, valve should drop out into your hand.

Step 12. *Only* if carburetor is excessively dirty or jets J (4-32) are obviously damaged or clogged is it necessary that these two jets be removed.

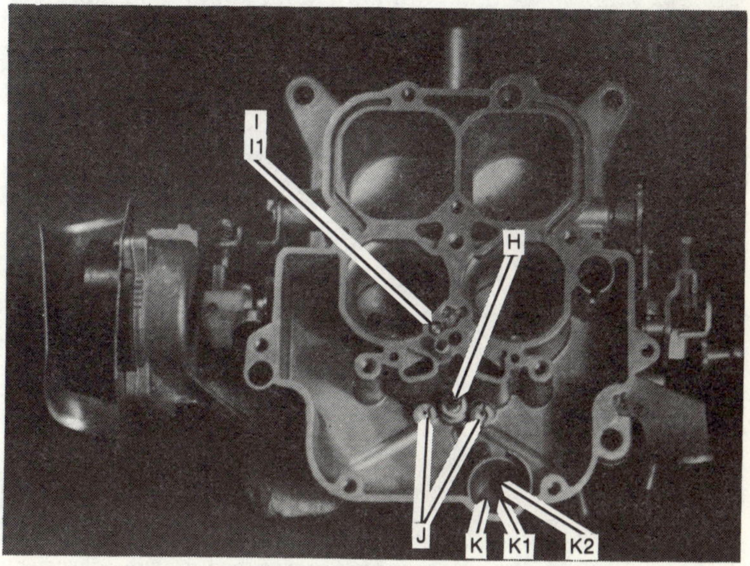

Fig. 4-32. Power valve, jets, accelerator-pump valve.

Step 13. Remove spring clip K (4-32) from bottom of accelerator pump well K1—use end of small screwdriver.

Step 14. Hold hand over well K1 (4-32), turn carburetor bowl over; check ball K2 should drop out of well into your hand.

Step 15. Place carburetor upside down before you as in 4-33.

Step 16. Use Phillips screwdriver; remove six screws L (4-33) from base; discard gasket; place bowl in bath.

Step 17. Remove choke housing A4 (4-33) by removing three screws A2 and three retainers A3. Discard gasket.

Step 18. Remove the two idle-mixture adjustment screws M (4-28) and springs M1. Place in bath.

Step 19. Place carburetor base in bath.

Carburetor disassembly is complete. Allow all parts placed in bath to be properly cleaned in accordance with directions on carburetor-cleaner container before beginning assembly. The remove all parts from cleaner, rinse with water, and allow to dry naturally. Do not use compressed air.

To Reassemble

Step 20. Install the idle-mixture adjustment screws M (4-33) and springs M1 into carburetor base. Tighten screws until they seat lightly, then back each one out about 1½ turns.

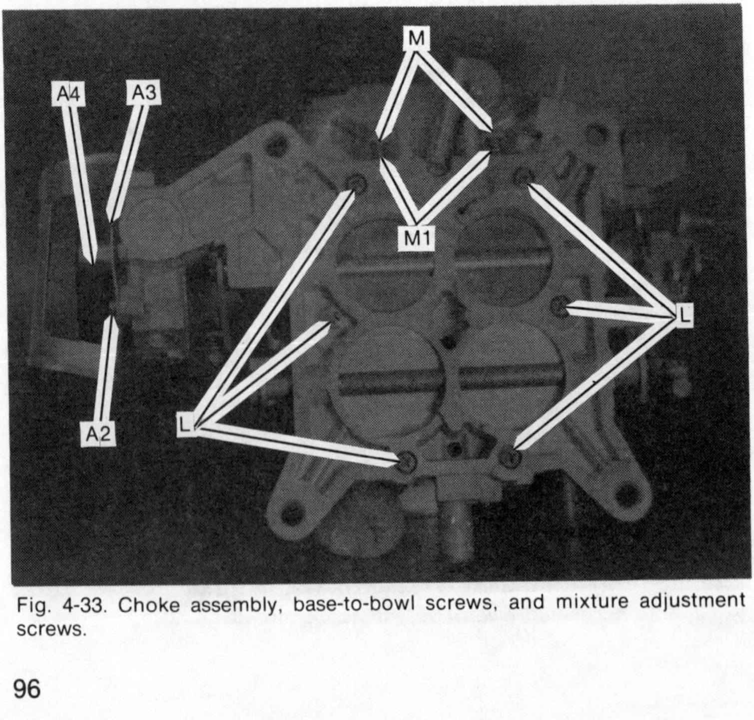

Fig. 4-33. Choke assembly, base-to-bowl screws, and mixture adjustment screws.

Step 21. Use new gasket; install carburetor bowl to base with the six screws L (4-33). Tighten screws firmly.

Step 22. Use new gasket. Install choke housing A4 (4-33) with retainers A3 and screws A2. Tighten screws finger-tight only.

Step 23. Rotate choke housing clockwise to engage spring and stop when mark on housing aligns with specified mark on metal base (see data sheet in parts kit for proper setting). Tighten screws A2 (4-33).

Step 24. Drop new steel check ball K2 (4-34) into well K1; install spring clip K into well K1.

Step 25. If a plastic retainer H1 is used on spring H2 of power valve H (4-34), replace it with metal one from parts kit. (If metal retainer is already in place, leave it there.)

Step 26. Use ⅜ in. socket to install power valve H (4-29). Tighten firmly.

Step 27. Drop a new accelerator pump discharge valve I (4-34) point first into well I1.

Step 28. Use broad-blade screwdriver; tighten the two jets J (4-34).

Step 29. Use new gasket. Install new needle-valve seat G1 (4-35). Tighten seat firmly.

Note: Some needle-valve seats in carburetors of this model must be tightened with ⅜ in. wrench, but most require a broad-blade screwdriver.

Step 30. Drop new needle valve G (4-35) into well G2.

Fig. 4-34. Accelerator-pump check valves, power valve, and related parts.

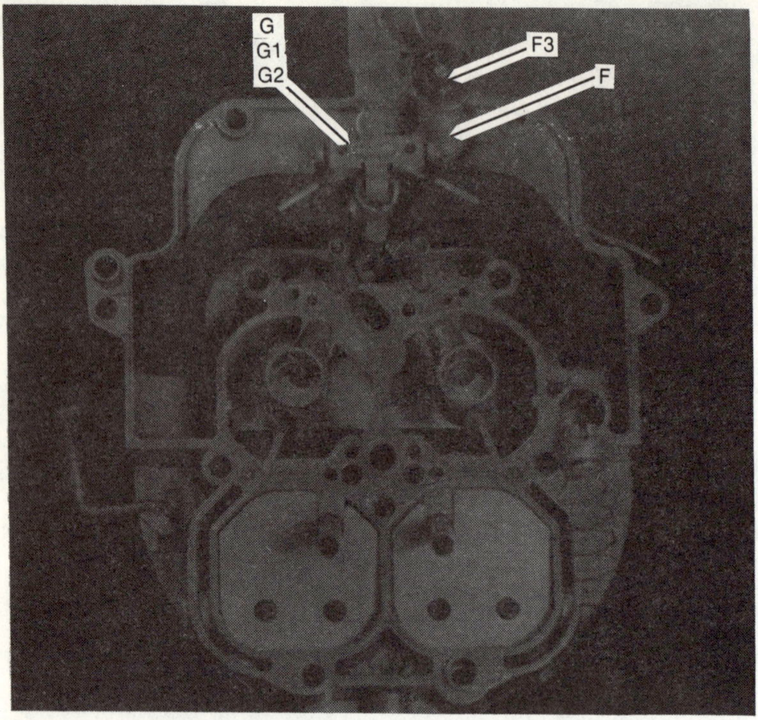

Fig. 4-35. Float assembly and accelerator pump.

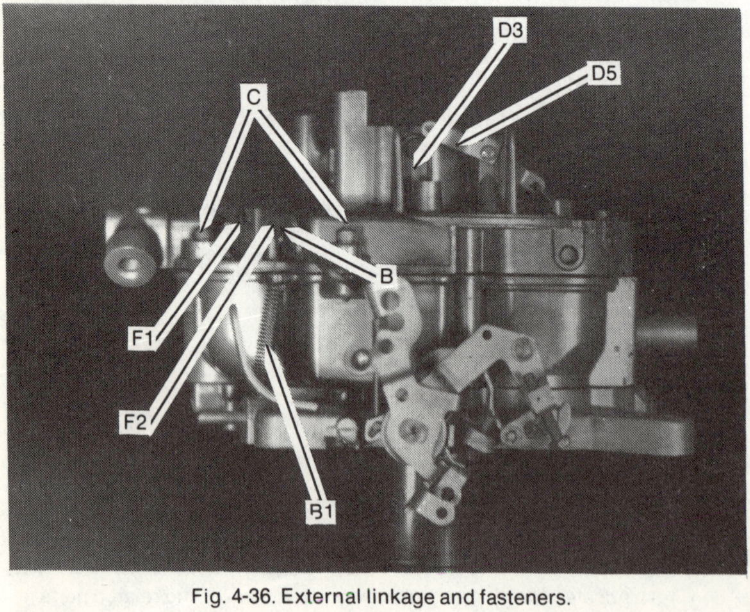

Fig. 4-36. External linkage and fasteners.

Step 31. Remove old rubber cup F3 (4-35) from end of accelerator pump assembly F, and install new cup from parts kit.

Step 32. Install accelerator pump assembly F (4-35) on bowl cover; attach pump-activating arm F2 (4-36) to accelerator pump plunger with pin F1.

Step 33. Install float; insert pin with head toward accelerator pump assembly.

Step 34. Make float adjustments in accordance with instructions in parts kit, and as shown in 4-37. Both the float level and the drop are important—make adjustments carefully.

Step 35. Place new gasket on bowl cover.

Step 36. Install air-valve piston assembly (refer to Fig. 4-27) into well. Insert lever D5 (4-36) into rod D3.

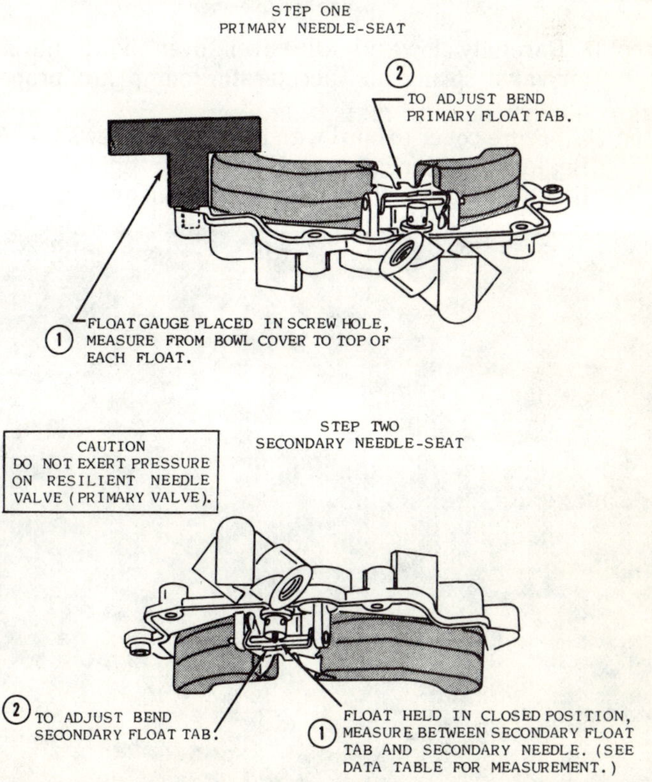

Fig. 4-37. Float adjustments.

Fig. 4-38. Choke linkage and idle-speed screw.

Step 37. Carefully lower bowl cover over bowl—be sure air-valve spring and accelerator pump are properly aligned.

Step 38. Secure cover to bowl with the eleven screws C (4-36). Begin at center and work outward; tighten screws firmly; retighten several times, until all are firm.

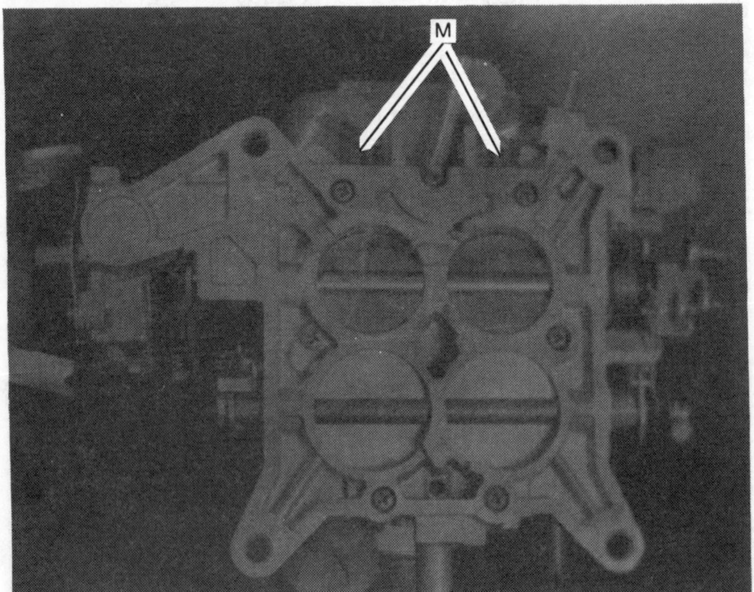

Fig. 4-39. Idle-mixture screws.

Step 39. Check air valve rod D3 (4-36) to be sure it is still attached to lever D5.

Step 40. Attach linkage B (4-36); attach spring B1.

Step 41. Attach choke linkage A (4-38) with pin clip A1.

Step 42. Mount carburetor on engine. Be sure fuel line and all linkages are properly attached.

Step 43. Start engine. Allow it to warm up until butterfly valves in carburetor barrels are fully open.

Step 44. Use screw M (4-38) to adjust idle to desired speed.

Step 45. Use screws M (4-39) to adjust idle mixture. Turn screws clockwise, one at a time, until engine begins to falter, then turn counterclockwise until it smooths out. Alternate adjustment from one screw to the other until engine sounds right.

Install the air cleaner. Be sure the element is clean.

Chapter 5

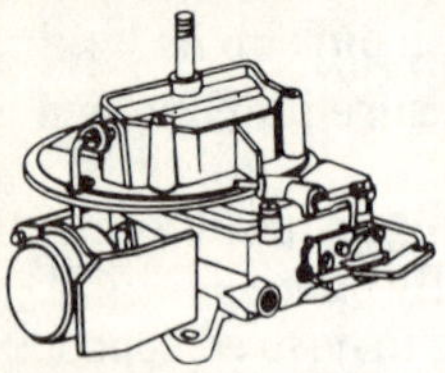

Holley

Holley makes a wide range of carburetors for domestic autos and trucks. While most production is limited to prosaic designs, the big four barrels have given the firm a reputation in racing circles second only to Weber. Some of the larger Holleys deliver more than 1000 cubic feet of air a minute.

Holley carburetors are mechanically simple and are remarkable for the clean bores, free of protrusions which would disrupt air flow.

HOLLEY ONE-BARRELS (MODELS 1904-1920)

Please refer to Chapter 1 for a brief description and discussion of carburetor cleaners and the container needed to make up the bath to be used for this task.

Only a few tools are needed for tuneup of this one-barrel carburetor; they are listed in Table 5-1.

Table 5-1. Tools Required for Holley Carburetor Overhaul.

Screwdriver with a broad blade (approx. $^7/_{16}$ in.).
Screwdriver with a narrow blade (approx. $^3/_{16}$ in.).
Screwdriver with medium Phillips head.
Needle-nosed pliers.
Open-end, box-end, or socket wrench of the proper size to remove the nuts or bolts that hold your carburetor to the manifold.
Note: Most air cleaners can be removed without tools or with only a pair of pliers.

Preliminary Instructions

Remove air cleaner.

Remove all dirt and grime from exterior of carburetor while carburetor is still mounted on engine; use kerosene and a paintbrush if necessary, but **do not use gasoline**.

Remove carburetor from engine.

For information on carburetor accessories, refer to Chapter 2. If you are not sure you can remember where and how the lines and linkages connect, make notes and sketches in the margin of this book as you remove the carburetor. Figure 5-1, an exploded view, is included to help you identify the various parts of your carburetor.

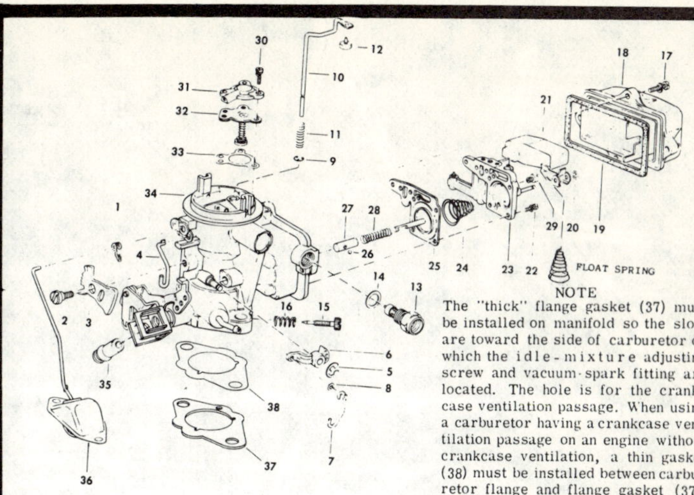

NOTE

The "thick" flange gasket (37) must be installed on manifold so the slots are toward the side of carburetor on which the idle-mixture adjusting screw and vacuum-spark fitting are located. The hole is for the crankcase ventilation passage. When using a carburetor having a crankcase ventilation passage on an engine without crankcase ventilation, a thin gasket (38) must be installed between carburetor flange and flange gasket (37). This thin gasket blocks the crankcase ventilation passage and prevents intake of air. The thin gasket must not be used on cars equipped with crankcase ventilation system.

Index No.	NOMENCLATURE	Index No.	NOMENCLATURE
1.	Choke rod retainer clip	20.	Float lever retainer ("E" washer)
2.	Fast-idle cam screw	21.	Float assembly
3.	Fast-idle cam	22.	Screw-and-lockwasher assembly
4.	Fast-idle rod	23.	Main-well and economizer body assy.
5.	Pump-lever retainer ("E" washer)	24.	Pump-diaphragm spring
6.	Pump lever	25.	Pump diaphragm assembly
7.	Pump link	26.	Pump push-rod sleeve ball
8.	Pump-link washer	27.	Pump push-rod sleeve
9.	Bowl vent rod retainer ("E" washer)	28.	Pump push-rod spring
10.	Bowl vent rod	29.	Main-metering jet
11.	Bowl vent rod spring	30.	Screw and lockwasher assembly
12.	Bowl-vent valve	31.	Economizer diaphragm cover
13.	Needle and seat assembly	32.	Economizer diaphragm assembly
14.	Needle-seat gasket	33.	Economizer diaphragm gasket
15.	Idle-mixture adjusting needle	34.	Body assembly
16.	Idle-adjusting needle spring	35.	Throttle-rod insulator bushing
17.	Screw-and-lockwasher assembly	36.	Automatic choke assy. (well type)
18.	Fuel bowl	37.	Manifold-flange gasket, thick (see "note", above)
19.	Fuel-bowl gasket	38.	Manifold flange gasket, thin (see "note", above)

Fig. 5-1. Exploded view of Holley one-barrel carburetor, Models 1904-1920.

When carburetor is removed from engine, follow tuneup steps listed below.

To Disassemble

Step 1. Remove hose A (5-2). Replacement hose will not be found in your parts kit; however, hose should be replaced if it shows any sign of poor fit, hardening, wear, or leakage. Ask for 5 in. of $^{3}/_{16}$ in. vacuum hose.

Step 2. Carefully note positions of the ends of spring B1 (5-2); on assembly this spring must be replaced in its original position. Remove unloader arm B and spring B1.

Step 3. Remove retainer clip C (5-3); remove linkage D and put it aside.

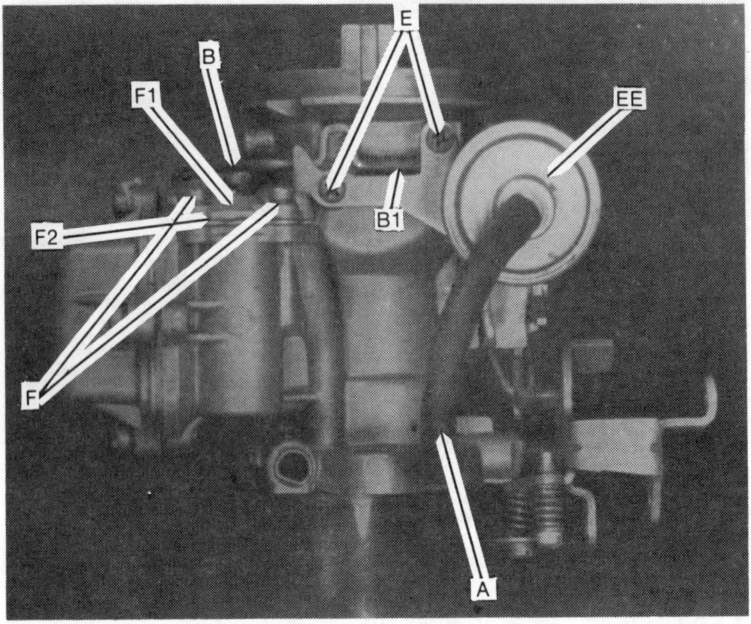

Fig. 5-2. External fasteners and fittings.

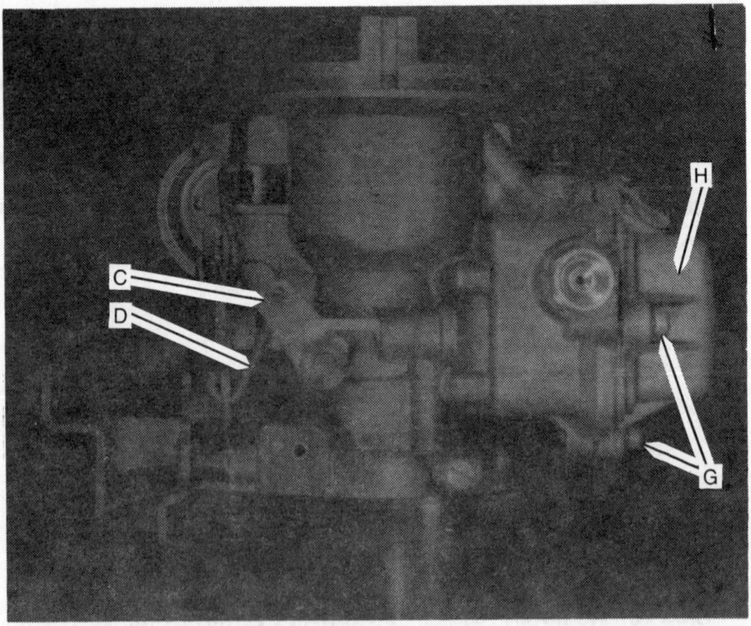

Fig. 5-3. Accelerator-pump linkage and float-chamber cover.

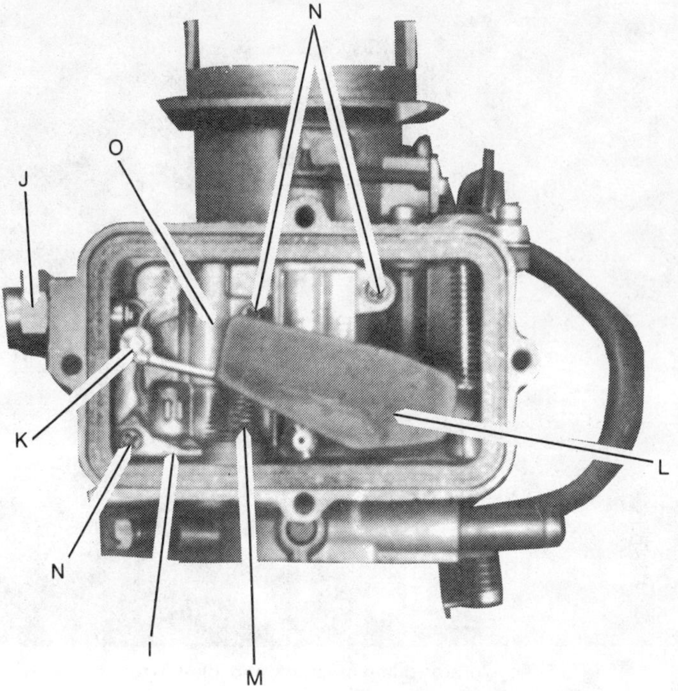

Fig. 5-4. Float and related parts.

Step 4. Remove the two screws E (5-2) which secure choke pull-off assembly EE (5-2); remove this assembly and put it aside.

Step 5. Remove the three screws F (5-2); lift off economizer-diaphragm assembly cover F1, and put it aside; discard diaphragm assembly F2 and gasket.

Step 6. Remove the four screws G (5-3); discard gasket; place float cover H in bath.

Step 7. Note carefully the position of baffle I (5-4) in fuel-bowl assembly; remove baffle and place in bath.

Step 8. Remove and discard needle-valve assembly J (5-4).

Step 9. Remove snapring K (5-4); lift out float assembly L (5-4), and cone-shaped helper spring M (5-4). Set all these aside.

Step 10. Remove the five small Phillips screws N (5-4) found in bowl behind float assembly. Notice that the lower left and upper right screws are shorter than the other three. Lift out the metering block assembly O held in place by these screws and place it in the bath.

Step 11. Remove cotter key P (5-5), and disconnect linkage Q at this point.

Fig. 5-5. Accelerator-pump assembly and idle-mixture screw.

Step 12. Lift out the accelerator-pump assembly R (5-5)—including spring R1 around shaft R2—from well R3 beneath metering block assembly O (removed in step 10). Set spring aside; discard assembly.

Step 13. Remove idle-mixture adjustment screw S (5-5) and spring T. Set them aside.

Step 14. Place carburetor body in bath.

Note: All metal parts indicated should now be in bath. Allow all parts placed in bath to be properly cleaned in accordance with directions on carburetor-cleaner container before beginning reassembly. Then remove all parts from cleaner, rinse with water, and allow to dry naturally. Do not use compressed air.

To Reassemble

Step 15. Place spring T over idle-mixture screw S (5-5); run screw assembly in until you feel it seat against bottom. Turn counterclockwise 1½ turns.

Step 16. Place spring R1 over shaft of new accelerator pump assembly; and insert shaft of assembly through well R3 at base of fuel bowl. Hold assembly all the way down against spring, and attach linkage Q; insert cotter key P (5-5).

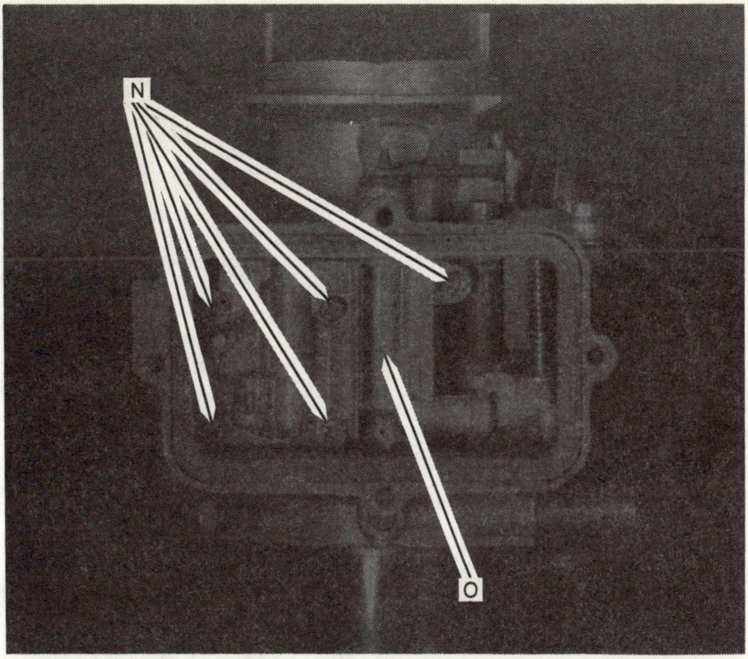

Fig. 5-6. Metering-block assembly.

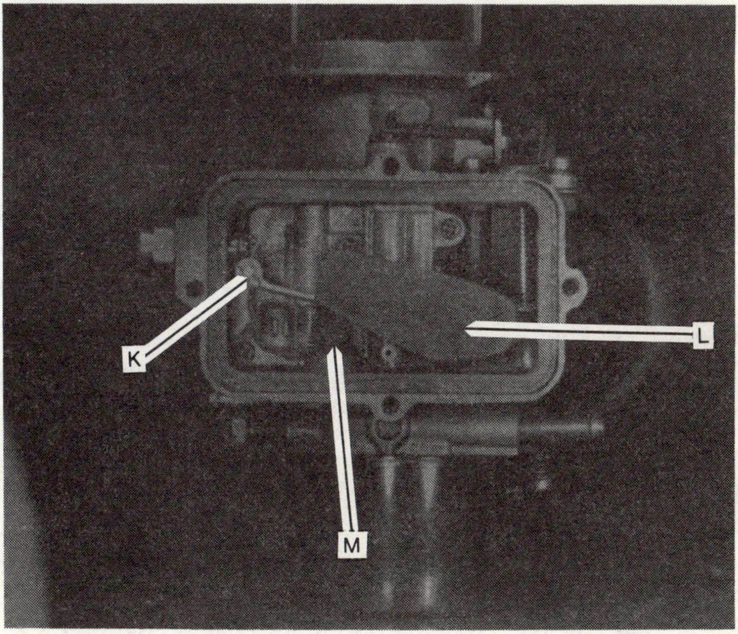

Fig. 5-7. Float assembly.

Fig. 5-8. Needle-and-seat assembly.

Step 17. Install metering block assembly O (5-6); insert the five Phillips screws N, and tighten them evenly and firmly.

Step 18. Replace float assembly L (5-7) and helper spring M.

Note: Narrow end of cone-shaped spring M points up, and *wire must be inserted in small hole in float arm.*

Step 19. Replace snapring K (5-7). Check float action.

Step 20. Use new gasket; install new needle-and-seat assembly J (5-8). Tighten firmly.

Step 21. Adjust float level as shown in 5-8. *Be accurate.*

Step 22. Use new gasket; replace bowl cover H (5-10); insert the four screws G; tighten them evenly and firmly.

Step 23. Use new gasket; install new economizer valve assembly F2 (5-11) **over** gasket; position cover F1. Be sure all holes are aligned; insert and tighten firmly the three screws F.

Step 24. Use new rubber mushroom-shaped valve V (5-11). Put a drop of oil on this valve before attempting to pull it through eye at end of unloader arm B.

Step 25. Replace unloader arm B and spring B1 (5-11). Be sure spring ends are in same position as when removed.

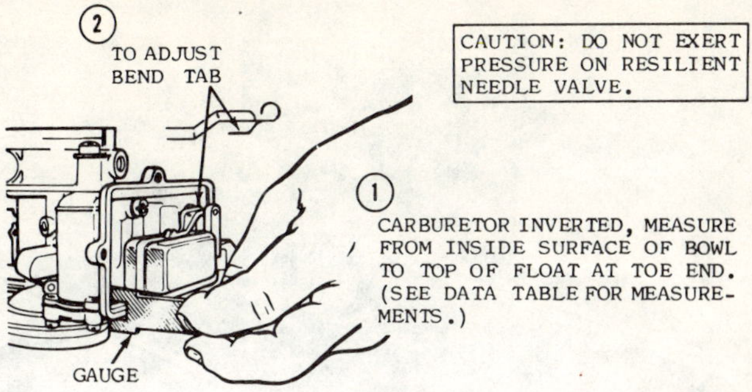

FLOAT LEVEL ADJUSTMENT

YEAR	MAKE	FLOAT LEVEL
1959-60	AMERICAN MTRS. 196" ENG.	9/32"
1961-62	AMERICAN MTRS. 196" ENG.	21/64"
1960-61	COMET 6 CYL. 144"-170" ENG.	9/32"
1959-60	EDSEL 6 CYL. 223" ENG.	9/32"
1960-61	FALCON 6 CYL. 144"-170" ENG.	9/32"
1952-64	FORD 6 CYL. 215"-223" ENG.	9/32"
1952-62	FORD TRUCK 6 CYL. 144"-170"-213"-215" ENG.	9/32"
1961-63	FORD TRUCK 6 CYL. 223"-262" ENG.	3/8 "
	GMC TRUCK 6 CYL. 248"-270" ENG.	3/8 "
	GMC TRUCK V6 305A ENG.	13/64"
ALL	INTERNATIONAL TRUCK 6 CYL	3/8 "
ALL	INTERNATIONAL HARVESTER SCOUT 4 CYL.	3/8 "
1961-62	MERCURY 223" ENG.	9/32"

DATA TABLE

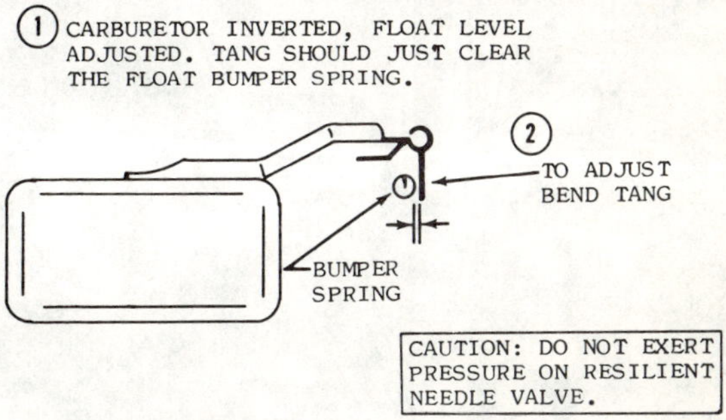

(CARBURETORS USING FLOAT BUMPER SPRING)
FLOAT DROP TANG ADJUSTMENT

Fig. 5-9. Float-level adjustment.

Step 26. Replace choke pull-off assembly EE (5-11); insert and tighten two screws E.

Step 27. Work the unloader arm B; it should move freely, and close with spring tension.

Step 28. Replace linkage D and retainer clip C (5-12).

Fig. 5-10. Float-chamber cover and screws.

Fig. 5-11. External fasteners and fittings.

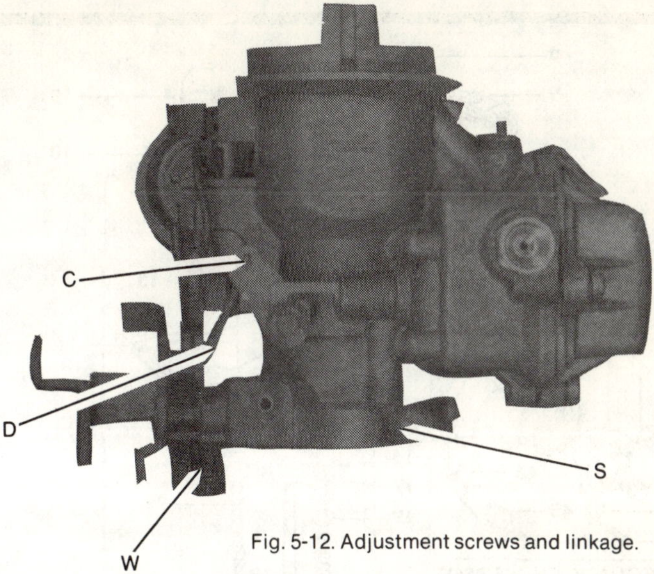

Fig. 5-12. Adjustment screws and linkage.

Step 29. Replace hose A (5-11).

Step 30. Install carburetor on car. Be sure all linkages and lines are properly connected.

Step 31. Start engine; allow to warm up until choke valve is fully open.

Step 32. Adjust idle speed adjustment screw W (5-12) until desired idle speed is achieved.

Step 33. Adjust idle mixture by turning idle-mixture adjustment screw S (5-12) clockwise until engine begins to run rough, then counterclockwise until idle becomes smooth.

Replace air cleaner on carburetor. Be sure element is clean.

HOLLEY ONE-BARREL CARBURETOR (MODEL 1909)

Please refer to Chapter 1 for a brief description and discussion of carburetor cleaners and the container needed to make up the bath to be used for this task. Tools required are listed in Table 5-1.

Preliminary Instructions

Remove air cleaner.

Remove all dirt and grime from exterior of carburetor while carburetor is still mounted on engine; use kerosene and a paintbrush if necessary, but **do not use gasoline.**

Remove carburetor from engine.

For information on carburetor accessories, refer to Chapter 2. If you are not sure you can remember where and how the lines

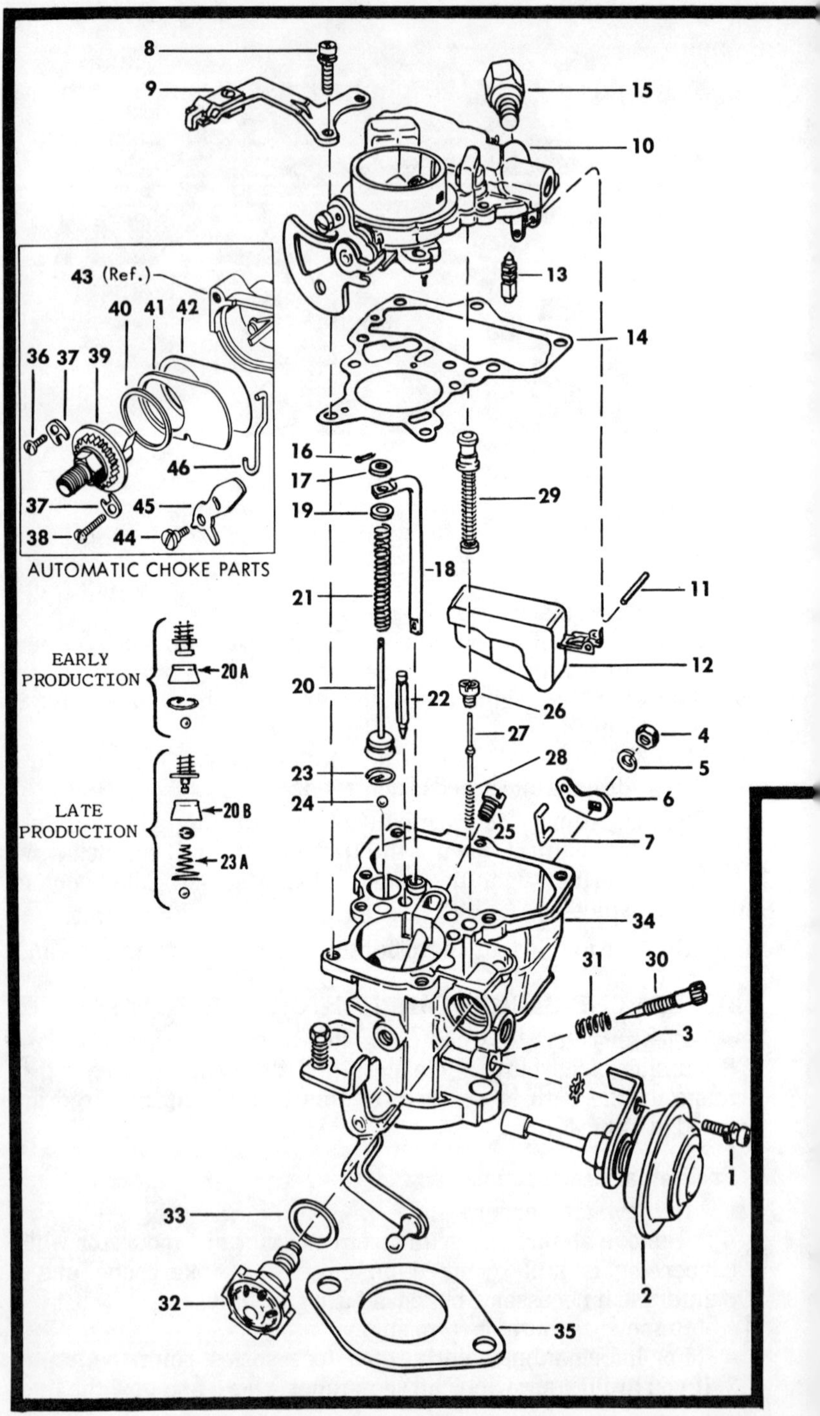

8
9
15
10
13
14
43 (Ref.)
40 41 42
36 37 39
46
37
45
38
44
AUTOMATIC CHOKE PARTS
16
17
19
29
18
21
11
EARLY PRODUCTION
20 A
12
20
22
26
27
4
28
23
5
24
6
LATE PRODUCTION
20 B
25
7
23 A
34
31
30
3
1
33
2
32
35

NOMENCLATURE

REF. NO.		REF. NO.	
1.	SCREW & LOCKWASHER-DASHPOT BRACKET	22.	NEEDLE-PUMP DISCHARGE
2.	DASHPOT ASSEMBLY	23.	RETAINER-PUMP INTAKE CHECK BALL
3.	LOCKWASHER	23A.	SPRING-PUMP RETURN & CHECK BALL RETAINER
4.	NUT-PUMP LEVER	24.	BALL-PUMP INLET CHECK
5.	LOCKWASHER-PUMP LEVER NUT	25.	JET-MAIN METERING
6.	LEVER-PUMP	26.	RETAINER-POWER VALVE
7.	LINK-PUMP	27.	VALVE-POWER
8.	SCREW & LOCKWASHER(8)-BOWL COVER	28.	SPRING-POWER VALVE
9.	BRACKET ASSY.-CHOKE CONTROL	29.	POWER VALVE PISTON ASSY.
10.	BOWL COVER ASSY.	30.	NEEDLE-IDLE ADJUSTING
11.	PIN-FLOAT HINGE	31.	SPRING-IDLE ADJ. NEEDLE
12.	FLOAT ASSEMBLY	32.	VALVE-SPARK
13.	NEEDLE-FUEL INLET	33.	GASKET-SPARK VALVE
14.	GASKET-BOWL COVER	34.	MAIN BODY ASSY.
15.	FITTING-FUEL INLET	35.	GASKET-FLANGE
16.	RETAINER-PUMP SPRING	36.	SCREW-STAT COVER RETAINER
17.	WASHER-PUMP PLUNGER	37.	RETAINER-STAT COVER
18.	ROD-PUMP OPERATING	38.	SCREW-STAT COVER RETAINER
19.	WASHER-PUMP PLUNGER	39.	STAT COVER ASSEMBLY
20.	PUMP PLUNGER ASSY.	40.	GASKET-STAT COVER
20A.	PUMP CUP-(FITS SNUG ON STEM)	41.	PLATE-CHOKE HOUSING
20B.	PUMP CUP-(FITS LOOSELY, HELD ON WITH E CLIP	42.	GASKET-CHOKE HSG. PLATE
21.	SPRING-PUMP	43.	BOWL COVER-(REF.)
		44.	SCREW-FAST IDLE CAM
		45.	CAM-FAST IDLE
		46.	LINK-CHOKE FAST IDLE

Fig. 5-13. Exploded view of Holley one-barrel carburetor, Model 1909.

and linkages connect, make notes and sketches in the margins of this book as you remove the carburetor. Figure 5-13, an exploded view, is included to help you identify the various parts of your carburetor.

When carburetor is removed from engine, follow tuneup steps listed below.

Disassembly

Step 1. Remove eight Phillips screws A (5-14); lift cover off bowl. Discard gasket.

Step 2. From cover, remove float-hinge pin B (5-15) and float C. Do *not* place in bath.

Step 3. Dump needle valve D (5-15) from seat D1. Discard needle valve.

Step 4. Use broad-blade screwdriver to remove needle-valve seat D1 (5-15). Discard seat and gasket.

Step 5. Place bowl cover in bath.

Step 6. Use adjustable wrench to remove power valve F (5-16) from well F1. Discard valve.

Fig. 5-14. Bowl-cover screws.

Step 7. Dump accelerator pump discharge needle valve G (5-16) from well G1. Place valve in bath.

Step 8. Use broad-blade screwdriver to remove jet H (5-16). Place jet in bath.

Step 9. Remove idle-mixture adjustment screw I (5-17) and spring J. Place both in bath.

Step 10. Use ⅜ in. or adjustable wrench to remove nut K (5-17) from shaft K1.

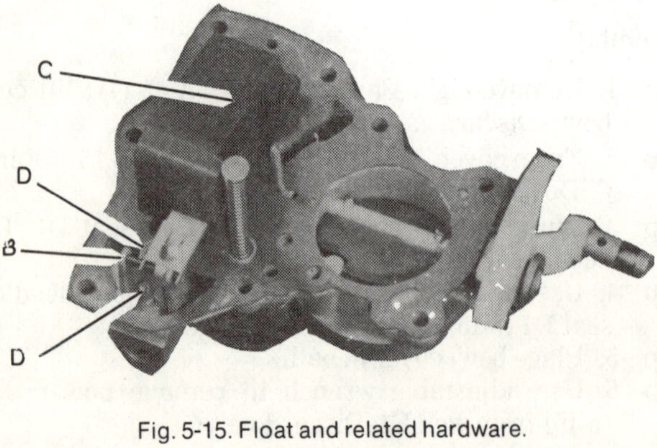

Fig. 5-15. Float and related hardware.

Fig. 5-16. Power valve and jet.

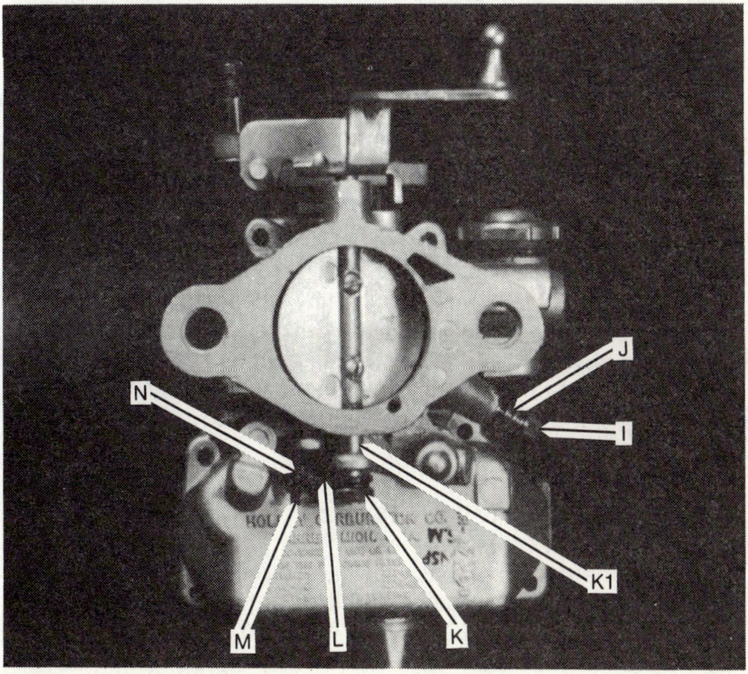

Fig. 5-17. Accelerator-pump linkage and idle-mixture screw.

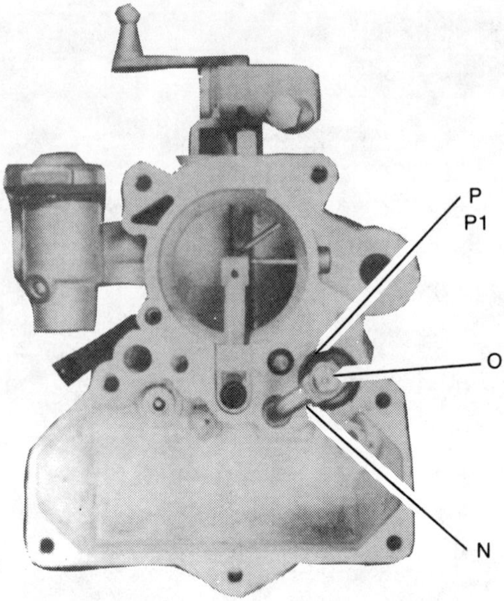

Fig. 5-18. Accelerator pump assembly.

Step 11. Slip lever L (5-17) off shaft K1; release it from linkage M.

Step 12. Slip linkage M (5-17) off lower end of accelerator pump activating rod N.

Step 13. Lift accelerator pump activating rod N (5-18) and accelerator pump assembly O from bowl; place this entire assembly in bath.

Step 14. Dump small steel check ball P (5-18) from well P1. Discard ball.

Step 15. Place carburetor bowl in bath.

Note: Disassembly is complete. Allow all parts in bath to be properly cleaned in accordance with the directions on the carburetor-cleaner container before beginning reassembly. Then remove all parts from bath, rinse them with water, and allow them to dry naturally. Do not use compressed air.

To Reassemble

Step 16. Drop new ball P (5-18) into well P1.

Step 17. From end of accelerator pump assembly O, remove cone-shaped coil spring—notice which way it points—then remove small *E* clip. Now, remove and discard the old Neoprene cup, and install the new one from your parts kit. Install *E* clip and cone-shaped coil spring.

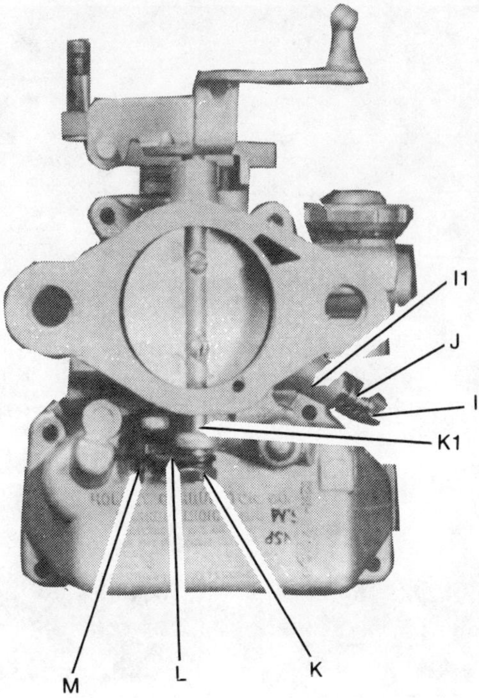

Fig. 5-19. Adjustment screw and accelerator-pump external linkage.

Step 18. Place accelerator pump assembly O (5-18) and rod N into position in carburetor bowl. Press down on assembly and attach linkage M (5-19) to N.

Step 19. Install lever L (5-19) into linkage M and over shaft K1. Secure in place with nut K.

Step 20. Place spring J (5-19) over screw I and insert into screw well I1. Run screw in until it seats lightly, then back it out 1½ turns.

Step 21. Use broad-blade screwdriver to tighten jet Q (5-20).

Step 22. Use broad-blade screwdriver to install jet H (5-20).

Step 23. Drop accelerator-pump discharge valve G (5-20), point first, into well G1.

Step 24. Use new gasket. Install new needle-valve seat D1 (5-21) into well D2.

Step 25. Drop new needle valve D (5-21) point-first into seat D1.

Step 26. Place new bowl-cover gasket into position on cover.

Step 27. Install float C (5-21) with hinge pin B.

Step 28. Adjust float level in accordance with instructions included in your parts kit.

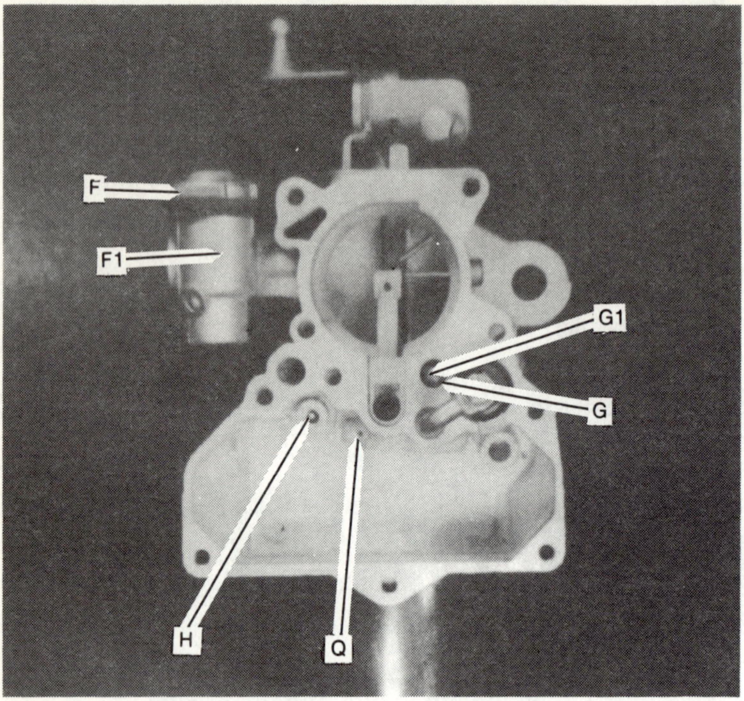

Fig. 5-20. Jets and accelerator-pump valves.

Step 29. Lower bowl cover into position on carburetor bowl. Secure with eight Phillips screws A (5-22). Tighten firmly.

Step 30. Install carburetor on engine. Be sure fuel line is attached tightly and that all linkages are properly connected.

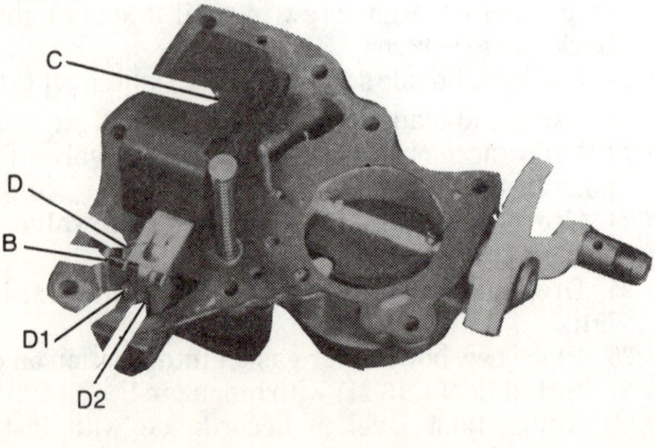

Fig. 5-21. Float and needle and seat.

Fig. 5-22. Float-cover screws, idle-speed, and mixture-adjustment screws.

Step 31. Start engine. Allow it to warm up until choke butterfly is in fully open position.

Step 32. Adjust idle to desired speed with screw R (5-22).

Step 33. Adjust idle mixture by turning screw I (5-22) clockwise until engine begins to falter, then counterclockwise until smooth idle is achieved.

Install air cleaner on carburetor. Be sure element is clean.

HOLLEY TWO-BARREL CARBURETOR (MODEL 2210)

Please refer to Chapter 1 for a brief description and discussion of carburetor cleaners and the solvent container needed to make up the bath to be used in this task. Tools are listed in Table 5-1.

Preliminary Instructions

Remove air cleaner.

Remove all dirt and grime from exterior of carburetor while carburetor is still mounted on engine; use kerosene and a paintbrush if necessary, but **do not use gasoline.**

Remove carburetor from engine.

For information on carburetor accessories, refer to Chapter 2. If you are not sure you can remember where and how the lines and linkages connect, make notes and sketches in the margin of this book as you remove the carburetor. Figure 5-23, an exploded view, is included to help you identify the various parts of your carburetor.

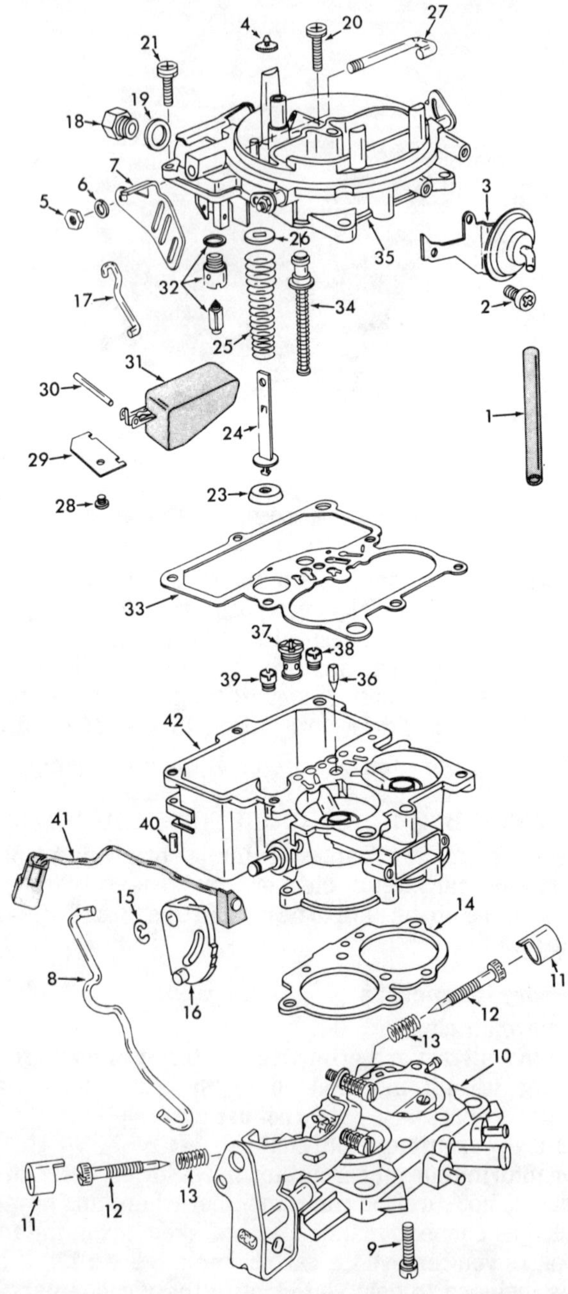
1
2
3
4
5
6
7
8
9
10
11
12
13
14
15
16
17
18
19
20
21
23
24
25
26
27
28
29
30
31
32
33
34
35
36
37
38
39
40
41
42

REF. NO.		REF. NO.	
1.	HOSE-CHOKE VACUUM	21.	SCREW & LOCKWASHER (7)-BOWL COVER
2.	SCREW (2)-CHOKE DIAPHRAGM BRACKET	22.	BOWL COVER ASSEMBLY
3.	CHOKE DIAPHRAGM AND LINK ASSY.	23.	CUP-PUMP
4.	VALVE-BOWL VENT	24.	PUMP STEM
5.	NUT-PUMP LEVER	25.	SPRING-PUMP DRIVE
6.	LOCKWASHER-PUMP LEVER	26.	WASHER-PUMP SPRING
7.	LEVER-PUMP	27.	PUMP LEVER SHAFT
8.	LINK-PUMP	28.	SCREW & LOCKWASHER-FUEL BOWL BAFFLE
9.	SCREW & LOCKWASHER (5)-THROTTLE BODY	29.	FUEL BOWL BAFFLE
10.	THROTTLE BODY ASSEMBLY	30.	PIN-FLOAT HINGE
11.	CAP-IDLE LIMITER*	31.	FLOAT & LEVER ASSEMBLY
12.	NEEDLE-IDLE ADJUSTING*	32.	NEEDLE, SEAT & GASKET ASSY.
13.	SPRING-IDLE ADJ. NEEDLE*	33.	GASKET-BOWL COVER
14.	GASKET-THROTTLE BODY	34.	PISTON ASSY.-POWER VALVE
15.	RETAINER-FAST IDLE CAM	35.	BOWL COVER
16.	CAM-FAST IDLE	36.	NEEDLE-PUMP DISCHARGE
17.	LINK-FAST IDLE CAM	37.	VALVE ASSY.-POWER
18.	FITTING-FUEL INLET	38.	JET-(63)-MAIN (CHOKE-SIDE)
19.	GASKET-FUEL INLET FITTING	39.	JET-(65)-MAIN (PUMP-SIDE)
20.	SCREW & LOCKWASHER (1)-BOWL COVER (LONG)	40.	PIN-WIRE RETAINER
		41.	STOP & CABLE ASSEMBLY
		42.	BOWL-FUEL

*DO NOT REMOVE UNLESS NEW CAPS ARE AVAILABLE

Fig. 5-23. Exploded view of Holley two-barrel carburetor, Model 2210.

When carburetor is removed from engine, follow tuneup steps listed below:

Step 1. Remove nut A (5-24).
Step 2. Remove lever B (5-24).
Step 3. Remove rod C (5-24).
Step 4. Remove hairpin clip D (5-24).

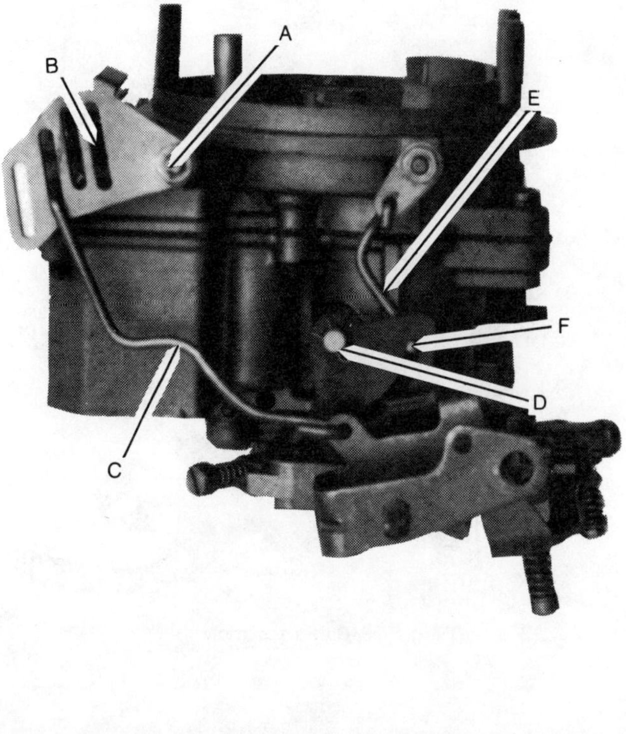

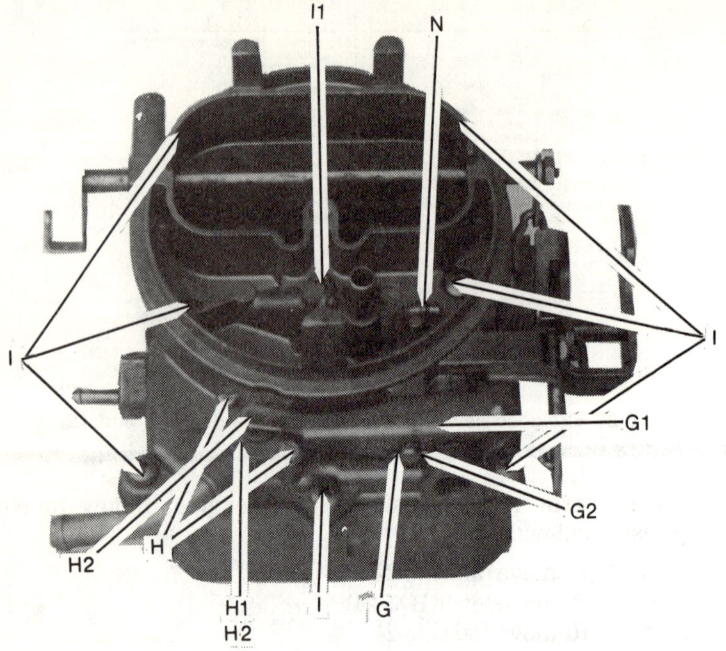

Fig. 5-25. Bowl cover-to-base screws and external fittings.

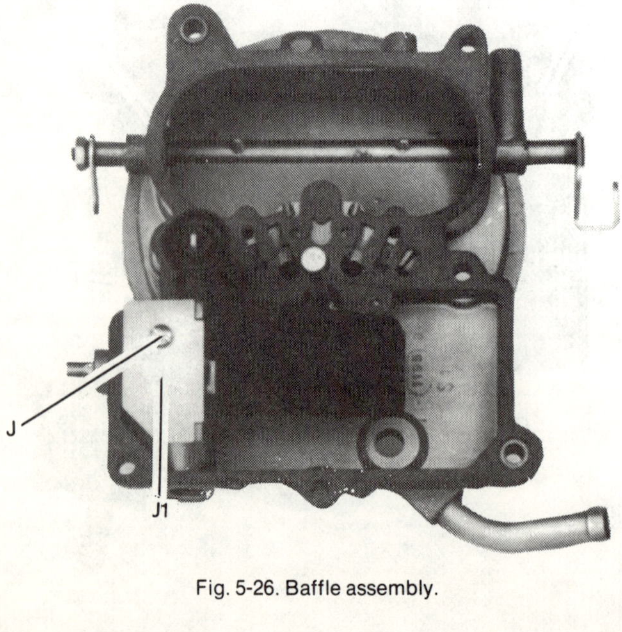

Fig. 5-26. Baffle assembly.

Step 5. Disconnect rod E (5-24) from fast-idle cam F.

Step 6. Remove hairpin clip G (5-25) from unloader lever G1.

Step 7. Lever G1 (5-25) is spring-loaded; remove lever G1 and small coil spring G2.

Step 8. Remove two screws H (5-25).

Step 9. Remove antipercolator unloader valve H1 (5-25) and, from beneath this valve, remove the fine-wound coil spring H2 (5-25).

Step 10. Remove the seven cover screws I (5-25) and the center screw I1. I1 is the larger of the two.

Step 11. Remove screw J (5-26).

Step 12. Remove baffle J1 (5-26); place baffle in bath.

Step 13. Remove hinge pin K (5-27); place pin in bath.

Step 14. Remove float K1 (5-27). Do *not* place float in bath.

Step 15. Remove needle valve L (5-27); discard needle valve.

Step 16. Use broad-blade screwdriver; remove needle-valve seat L1 (5-27); discard seat.

Step 17. Use thumb and finger to press down on accelerator pump assembly M (5-27); release shaft at N (5-25) by turning spring cover until notch lines up with shaft end; place entire assembly in bath.

Step 18. Remove and discard bowl-cover gasket; place bowl cover in bath.

Step 19. Remove accelerator-discharge needle valve O (5-28); place valve in bath.

Step 20. Remove the two metering jets P (5-28); place jets in bath.

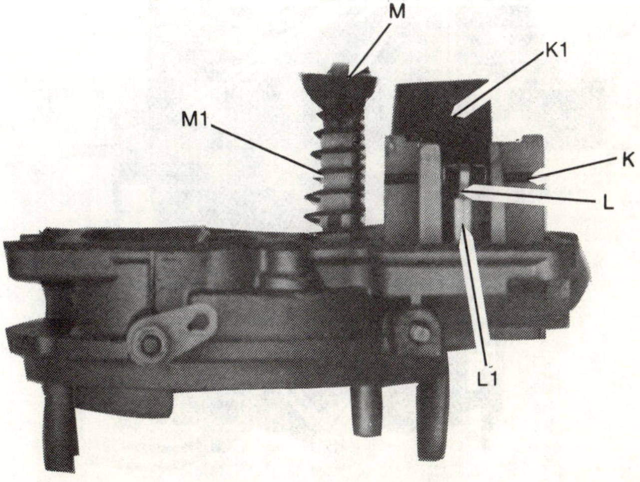

Fig. 5-27. Float and accelerator pump assembly.

Fig. 5-28. Discharge needle-valve, metering jets, and power jet.

Step 21. Remove power jet Q (5-28); place jet in bath.
Step 22. Remove the five screws R (5-29).
Step 23. Separate bowl from base; discard gasket; place bowl in bath.

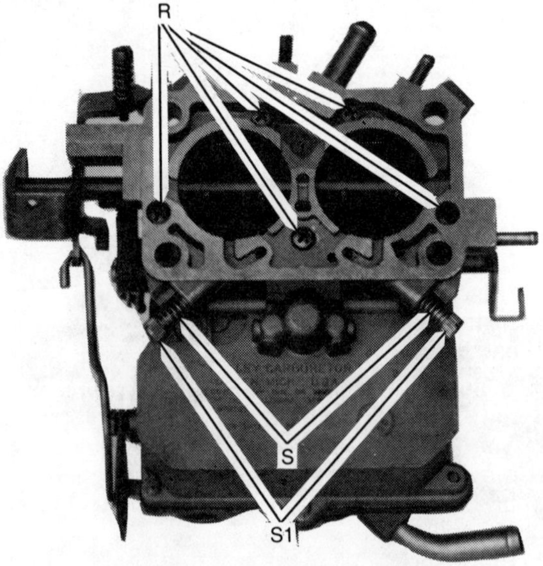

Fig. 5-29. Idle-mixture and base-to-bowl screws.

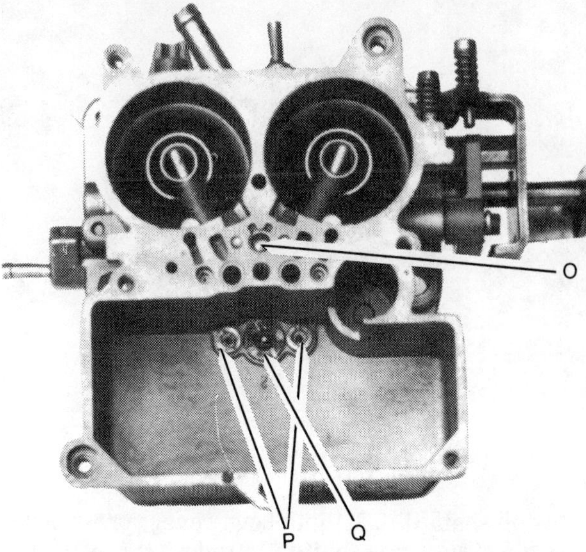

Fig. 5-30. Discharge needle valve, metering jets, and power jet.

Step 24. Remove the two idle-mixture screws R (5-29) and springs R1 from base; place screws and springs in bath.

Step 25. Place base in bath.

Note: Allow all parts placed in bath to be properly cleaned in accordance with directions on carburetor-cleaner container before beginning reassembly. Then, remove all parts from cleaner, rinse with water, and allow to dry naturally. Do not use compressed air.

To Reassemble

Step 26. Install the two screws S (5-29) and springs S1 on base. Run screws all the way in—finger-tight—then back out 1½ turns.

Step 27. Use new gasket. Position bowl on base; insert and tighten the five screws R (5-29).

Step 28. Install power jet Q (5-30); tighten firmly with broad-blade screwdriver.

Step 31. Install new needle-valve seat L1 (5-31) over new gasket; tighten firmly with broad-blade screwdriver.

Step 32. Your parts kit may contain only new cup M2 (5-31), or it may contain cup M2 assembled on a new shaft M. If your kit contains only cup M2, remove old cup from shaft, and install new cup.

Step 33. Place spring M1 over shaft M (5-31); the round, flat end of the spring must be against cup M2.

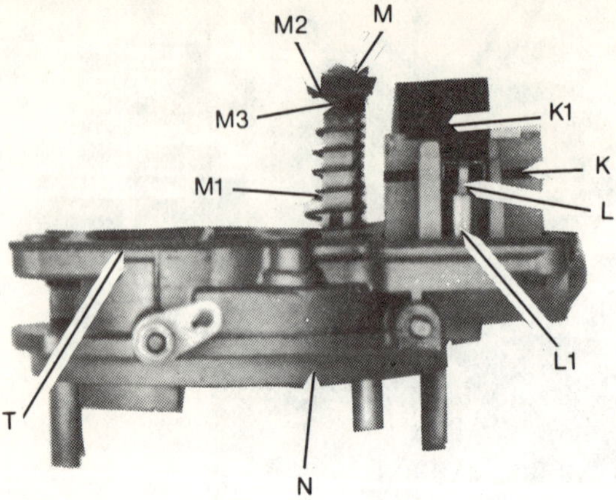

Fig. 5-31. Float and accelerator pump assembly.

Step 34. Install shaft M (5-31) into bowl cover; press down on assembly so that shaft protrudes through at N; assemble notched coupler on shaft end.

Step 35. Use new gasket. Position gasket on bowl cover at point T (5-31).

Step 36. Install new needle valve L (5-31) into seat L1, point first.

Step 37. Place float K1 (5-31) into position and install hinge pin K.

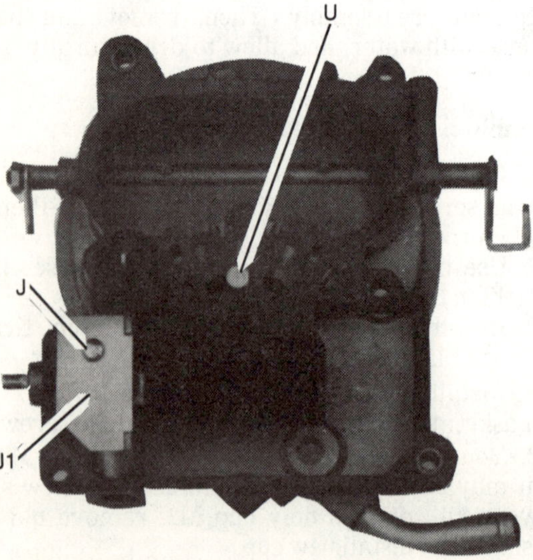

Fig. 5-32. Baffle, baffle screw, and vacuum piston assembly.

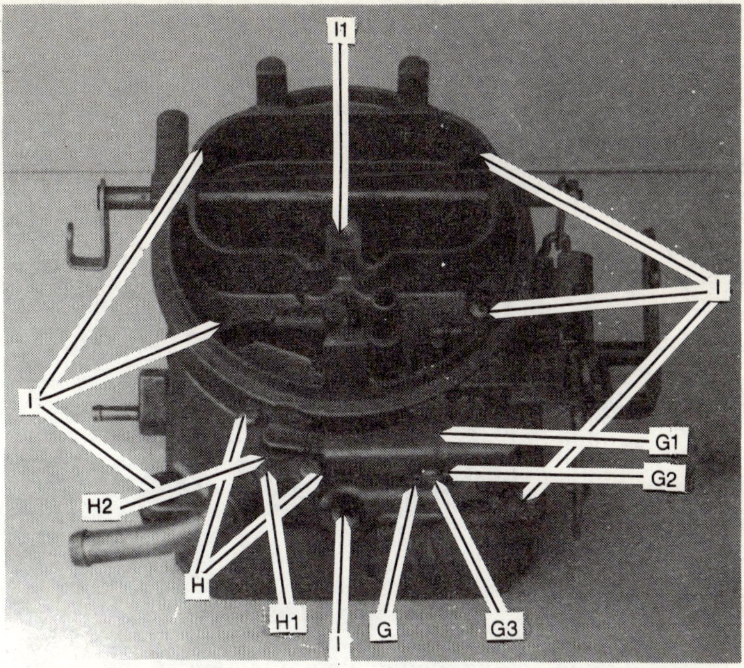

Fig. 5-33. Bowl cover-to-base screws, external fittings, lever, and spring.

Note: This pin is made of plastic or nylon and can be easily broken; once you have it started, it may be necessary to tap it gently with a screwdriver handle to drive it home.

Step 38. Adjust float level in accordance with instruction sheet included in your repair parts kit.

Step 39. Position baffle J1 properly and install it with screw J (5-32).

Step 40. Alternately press down and release the vacuum-piston assembly U (5-32) several times to be sure it is working freely. (If not, this assembly must be replaced with a new one.)

Step 41. Carefully place bowl cover into position on bowl; install and tighten the seven screws I (5-33).

Step 42. Install and tighten screw I1 (5-33). Remember, I1 is longer than those designated I.

Step 43. Place the fine-wound coil spring H2 (5-33) into well beneath H1.

Step 44. Install antipercolator unloader valve H1 (5-33) with the two screws H. Tighten screws.

Step 45. Install activating arm G1 (5-33), and coil spring G2. As you slip the arm on shaft G3, slip spring on shaft also, between the two sides of the arm. Be sure,

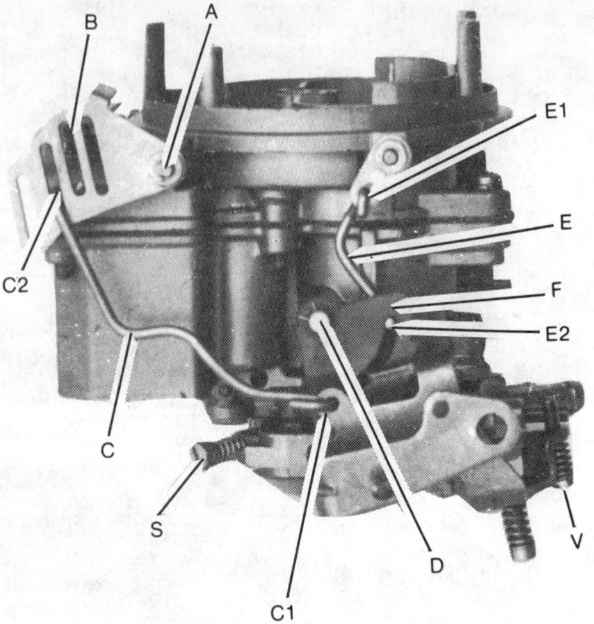

Fig. 5-34. Linkage, idle-speed screw and mixture-adjustment screws.

however, that both ends of the spring are pointed toward G2 as in the illustration (5-33). If spring is properly installed, arm will press down on valve H1.

Step 46. Replace hairpin clip G (5-33).

Step 47. Attach rod E (5-34) at E1 and E2.

Step 48. Replace hairpin clip D (5-34).

Step 49. Attach rod C (5-34) at C1 and to arm B at C2.

Step 50. If provided, install tension spring on shaft at A (5-34); then slide arm B on shaft.

Step 51. Install lockwasher and nut on shaft at A (5-34); tighten nut with ⅜ in. wrench.

Step 52. Mount carburetor on engine. Be sure all lines and linkages are properly connected.

Step 53. Start engine. Allow engine to warm up until choke valves are fully open.

Step 54. Adjust idle-speed adjustment screw V (5-34) for optimum engine speed.

Step 55. Adjust idle mixture with idle-mixture screws S (5-34); run one screw in until idle becomes rough, then back it out until smoothest engine idle is obtained. Repeat procedure with other screw; it may be necessary to go from one screw to the other several times before best idle is obtained.

Replace air cleaner. Be sure element is clean.

HOLLEY FOUR-BARREL CARBURETOR (MODEL 3310)

Please refer to Chapter 1 for a brief description and discussion of carburetor cleaners and the container needed to make up the bath to be used for this task. Tools are listed in Table 5-1.

Preliminary Instructions

This set of instructions is based upon the Holley Model 3310 (780 cubic feet per minute) dual-feed carburetor with vacuum secondary and automatic choke. If your Holley does not have the vacuum secondary and automatic choke, ignore references to these features in instructions. All else is the same on all models of this carburetor.

Note: Primary side of carburetor can be identified easily as the side with butterfly valve in barrels.

Remove air cleaner.

Remove all dirt and grime from exterior of carburetor while carburetor is still mounted on engine; use kerosene and a paintbrush if necessary, but **do not use gasoline.**

Remove carburetor from engine.

For information on carburetor accessories, refer to Chapter 2. If you are not sure you can remember where and how the lines and linkages connect, make notes and sketches in the margin of this book as you remove the carburetor. Figure 5-35, an exploded view, is included to help identify the various parts of your carburetor.

When carburetor is removed from engine, follow tuneup steps listed below:

To Disassemble

Step 1. Remove the four screws A holding accelerator pump housing AA (5-36).

Step 2. Discard gasket-and-diaphragm assembly B (5-36); place compression spring C in bath. Place pump housing AA in bath.

Step 3. Note check ball D at bottom of diaphragm bowl BB (5-36). It is not necessary to remove this ball, but it should be cleaned with care while carburetor bowl is in bath.

Step 4. Remove the two slotted screws E (5-37). Put them aside.

Step 5. Remove the four screws F (5-37) that hold the primary fuel bowl assembly G. Remove fuel-bowl assembly. Discard gasket.

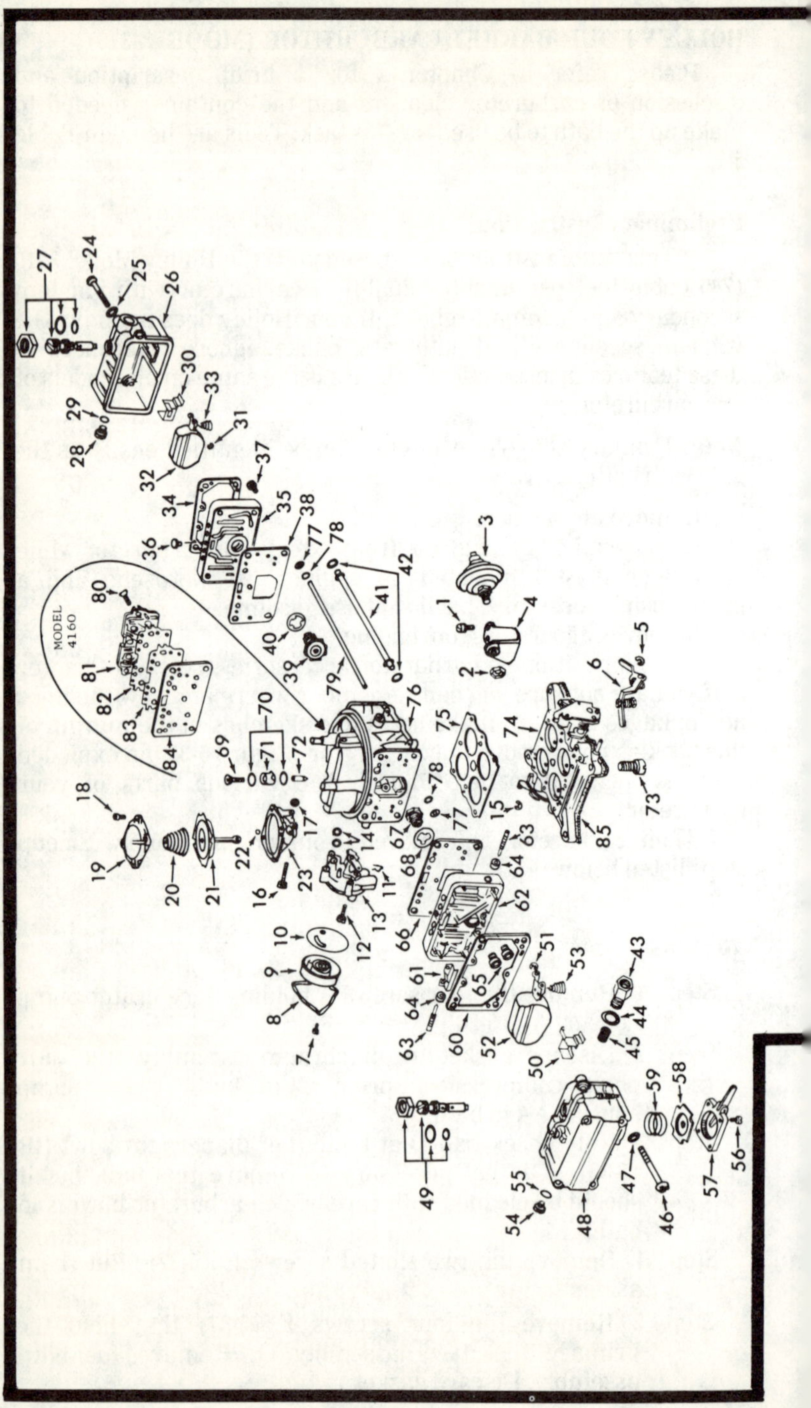
MODEL
4160

REF. NO.		REF. NO.	
1.	SCREW-DASHPOT BRACKET	23.	HOUSING-SEC. DIAPHRAGM
2.	LOCKNUT-DASHPOT	24.	SCREW-SEC. FUEL BOWL (4)
3.	DASHPOT	25.	GASKET-FUEL BOWL SCREW (4)
4.	BRACKET-DASHPOT	26.	FUEL BOWL-SEC.
5.	RETAINER-PUMP LEVER	27.	NEEDLE SEAT, "O" RINGS AND NUT ASSEMBLY. SEC.
6.	LEVER-PUMP	28.	PLUG-SIGHT LEVEL
7.	SCREW-STAT COVER (3)	29.	GASKET-SIGHT LEVEL PLUG
8.	STAT COVER CLAMP	30.	BAFFLE PLATE-SEC.
9.	STAT COVER	31.	RETAINER-SEC. FLOAT
10.	GASKET-STAT COVER	32.	FLOAT ASSEMBLY-SEC.
11.	RETAINER-CHOKE ROD	33.	SPRING-SEC. FLOAT
12.	SCREW & LOCKWASHER-CHOKE HOUSING (3)	34.	GASKET-SEC. FUEL BOWL
13.	CHOKE-HOUSING ASSEMBLY	35.	SEC. METERING BLOCK
14.	GASKETS-CHOKE HOUSING (2)	36.	VENT VALVE-SEC.
15.	RETAINER-SEC. DIAPHRAGM LINK	37.	JETS-SEC. MAIN (2)
16.	SCREW & LOCKWASHER-DIAPHRAGM HOUSING (3)	38.	GASKET-SEC. METERING BLOCK
17.	GASKET-SEC. DIAPHRAGM HOUSING	39.	POWER VALVE-SEC. (TRUCKS OVERRUN VALVE STAMPED 215)
18.	SCREW & LOCKWASHER-DIAPHRAGM COVER (4)	40.	GASKET-POWER VALVE
		41.	FUEL LINE TUBE
19.	COVER-DIAPHRAGM HOUSING	42.	"O" RINGS-FUEL LINE TUBE (2)
20.	SPRING-SEC. DIAPHRAGM	43.	FITTING-FUEL INLET
21.	DIAPHRAGM-SECONDARY	44.	GASKET-FUEL INLET
22.	BALL-SEC. DIAPHRAGM CHECK	45.	SCREEN-FUEL INLET

REF. NO.		REF. NO.	
46.	SCREW-FUEL BOWL PRI. (4)	68.	GASKET-PRI. POWER VALVE
47.	GASKET-FUEL BOWL SCREW (4)	69.	SCREW-PUMP DISCHARGE NOZZLE
48.	FUEL BOWL ASSEMBLY PRI.	70.	GASKETS-PUMP DISCHARGE NOZZLE (2)
49.	NEEDLE SEAT, "O" RINGS AND NUT ASSEMBLY. PRI.	71.	PUMP-DISCHARGE NOZZLE
50.	BAFFLE PLATE-PRI.	72.	PUMP-DISCHARGE NEEDLE VALVE (OR BALL AND WEIGHT)
51.	RETAINER-PRI. FLOAT		
52.	FLOAT ASSEMBLY-PRI.	73.	SCREW & LOCKWASHER-THROTTLE BODY (8)
53.	SPRING-PRI. FLOAT		
54.	PLUG-SIGHT LEVEL	74.	THROTTLE BODY ASSEMBLY
55.	GASKET-SIGHT LEVEL PLUG	75.	GASKET-THROTTLE BODY
56.	SCREW & LOCKWASHER-PUMP COVER (4)	76.	MAIN BODY ASSEMBLY
		77.	WASHER-BALANCE TUBE (2) 4160
57.	PUMP COVER ASSEMBLY	78.	"O" RING SEAL-BALANCE TUBE (2)
58.	DIAPHRAGM-PUMP	79.	BALANCE TUBE 4160
59.	SPRING-DIAPHRAGM RETURN	80.	SCREW-SEC. METERING BODY (6) 4160
60.	GASKET-PRI. FUEL BOWL		
61.	BAFFLE-PRI. METERING BODY VENT	81.	METERING BODY-SEC. 4160
		82.	GASKET-SEC. METERING BODY PLATE 4160
62.	PRI. METERING BLOCK		
63.	IDLE ADJUSTING NEEDLES (2)	83.	PLATE-SEC. METERING BODY 4160
64.	SEALS-IDLE NEEDLE (2)	84.	GASKET-SEC. METERING BODY 4160
65.	JETS-PRI. MAIN (2)	85.	MANIFOLD FLANGE GASKET
66.	GASKET-PRI. METERING BLOCK		
67.	POWER VALVE-PRI.		

Fig. 5-35. Exploded view of Holley four-barrel carburetor, Model 3310.

Step 6. Remove needle-valve assembly H (5-37) from beneath screw E by turning nut counterclockwise. Discard this entire assembly and gasket.

Step 7. From inside back of primary fuel bowl, remove the two screws I holding float assembly II (5-38)—pull the assembly straight out. Put float assembly aside. Place fuel bowl in bath.

Step 8. Remove primary fuel metering block J (5-39). There are no additional screws holding this block to body of carburetor, so if it adheres to body, pry it free with screwdriver blade. Discard gasket.

Fig. 5-36. Accelerator pump parts.

Fig. 5-37. Float-chamber covers and needle-valve assemblies.

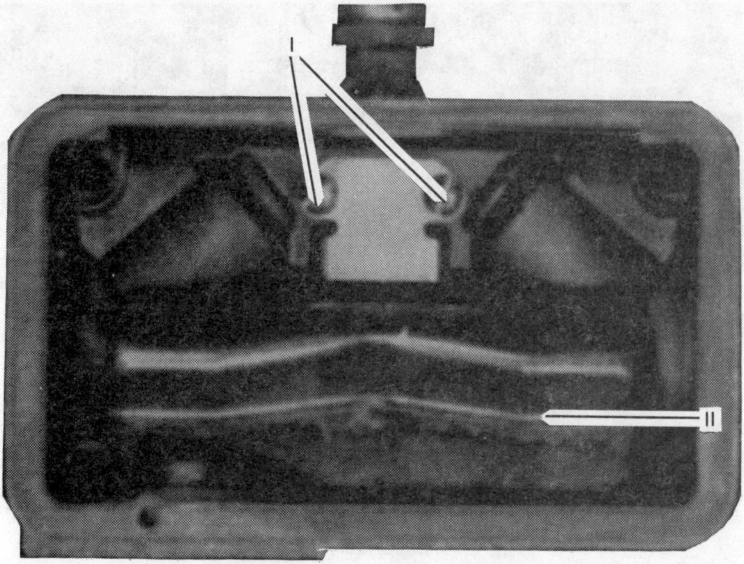

Fig. 5-38. Primary float bowl and related parts.

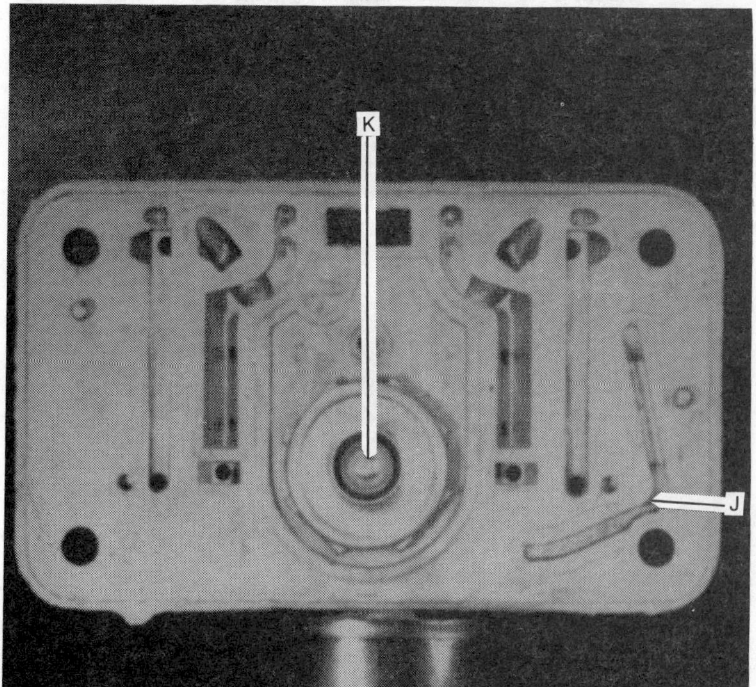

Fig. 5-39. Primary fuel-metering block and power valve.

Fig. 5-40. Secondary float cover.

Step 9. Remove power valve K (5-39) from metering block by turning it counterclockwise. Discard power valve. Place metering block in bath.

Step 10. Remove the four screws L (5-40) that hold the secondary fuel bowl assembly M. Remove fuel bowl assembly. Discard gasket.

Step 11. Remove needle-valve assembly HH from beneath screw E by turning nut counterclockwise (5-37). Discard the entire assembly and gasket.

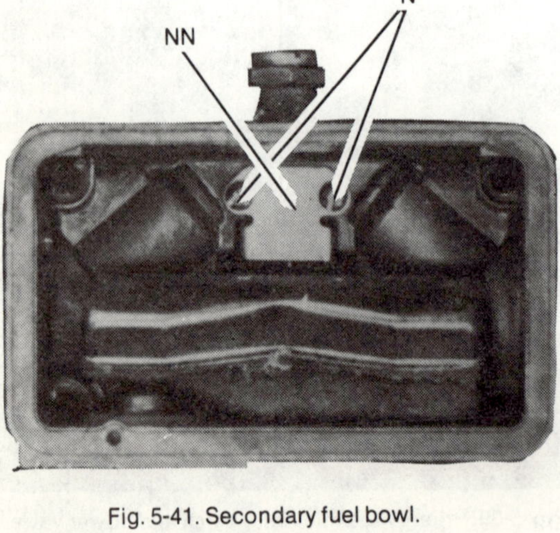

Fig. 5-41. Secondary fuel bowl.

Fig. 5-42. Secondary fuel-metering block.

Step 12. From inside back of secondary fuel bowl, remove the two screws N holding float assembly NN (5-41)—pull the assembly straight out. Put float aside. Place fuel bowl in bath.

Step 13. Remove secondary fuel metering block O (5-42).

Step 14. Remove the three screws P (5-43); remove retainer ring Q, Bakelite choke cover R, and put them aside.

Step 15. Remove hairpin clip S (5-43) from lower end of choke rod T.

Step 16. Remove three screws U (5-43) that hold choke assembly RR to carburetor body.

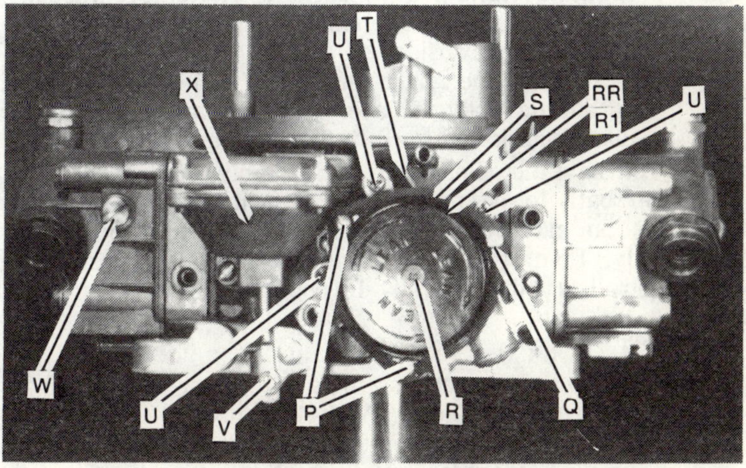

Fig. 5-43. Choke, vacuum secondary, assorted fasteners, and linkages.

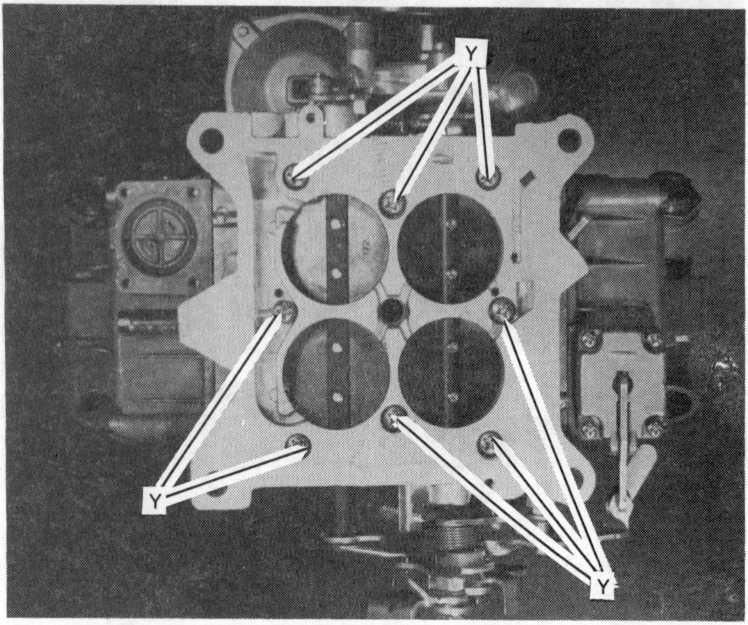

Fig. 5-44. Base-to-bowl screws.

Note: Notice relationship of choke rod to the brown plastic cam R1. Remove choke assembly RR and at the same time, disengage choke rod T. Discard the small, round cork gasket. Place choke assembly in bath.

Step 17. Remove snapring V (5-43).

Step 18. Remove the three screws W that hold vacuum secondary X to body of carburetor. Remove vacuum secondary; discard the small, round cork gasket; put entire assembly aside.

Step 19. Remove the eight screws Y (5-44) that hold carburetor body to base; separate body from base; discard gasket; place body and base in bath.

Note: You have completed the disassembly. Allow all parts placed in bath to be properly cleaned in accordance with directions on carburetor-cleaner container before beginning reassembly. Then remove all parts from cleaner, rinse with water, and allow to dry naturally. Do not use compressed air.

To Reassemble

Step 20. Use new gasket; assemble carburetor body to base; tighten all eight screws Y (5-44) firmly.

Step 21. Use new, small, round cork gasket; attach the vacuum secondary X to carburetor body. Tighten screws W.

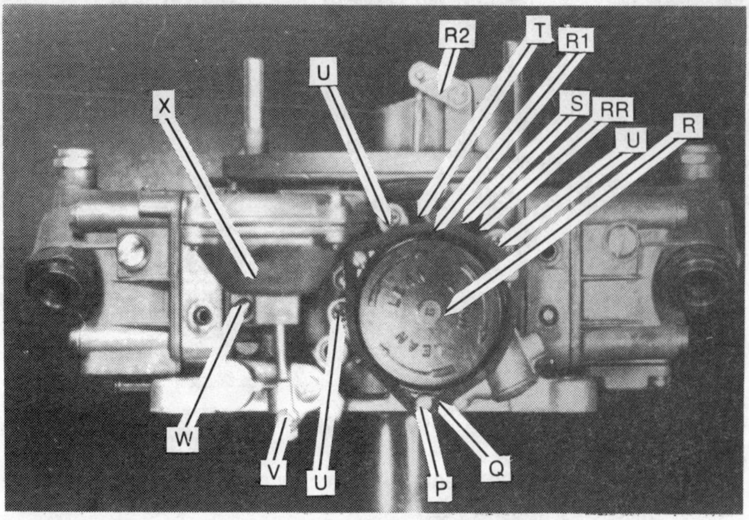

Fig. 5-45. External fittings.

Note: Use a tiny amount of heavy lube to hold gasket in place during assembly.

Step 22. Replace snapring V (5-45).

Step 23. Use new, small, round cork gasket; replace choke assembly RR on carburetor body—do not tighten screws U; be sure rod T (5-45) is properly inserted in arm R2, and in proper relation to the brown plastic cam R1. Tighten screws U. Be sure rod T works freely.

Step 24. Replace hairpin clip S on rod T (5-45).

Step 25. Replace Bakelite cover R (5-45) on choke assembly RR; replace retainer ring Q; insert and tighten three screws P.

Step 26. Replace float assembly NN (5-46) in secondary fuel bowl M; insert and tighten two screws N; work float to be sure it moves freely.

Note: On this carburetor, the float levels are adjusted externally after the carburetor is installed on car.

Step 27. Use new gasket; place secondary metering block O (5-47) in proper position against carburetor body, with the two metering jets OO (main jets) toward bottom of carburetor and facing out toward bowl.

Note: Be sure these jets are firmly seated; use good broad-blade screwdriver to tighten them.

Step 28. Use new gasket; place fuel bowl in proper position against metering block; insert screws L (5-48), using new rubber *O* rings L1; be sure all gaskets stay in place,

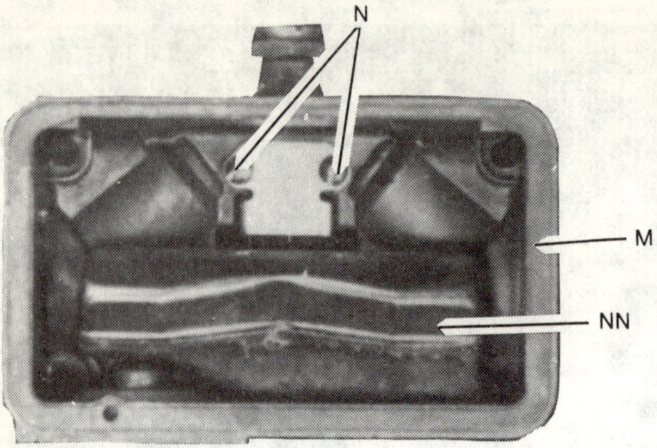

Fig. 5-46. Secondary float assembly.

and use good broad-blade screwdriver to tighten screws evenly and very firmly.

Step 29. Replace float assembly I1 (5-49) in primary fuel bowl; insert and tighten the two screws I; work float to be sure it works freely.

Step 30. Tighten main jets JJ (5-50) in primary metering block J.

Step 31. Use new gasket; install new power valve K in primary metering block J.

Step 32. Use new gasket; position metering block J (5-50) against carburetor body, with main jets JJ down and out.

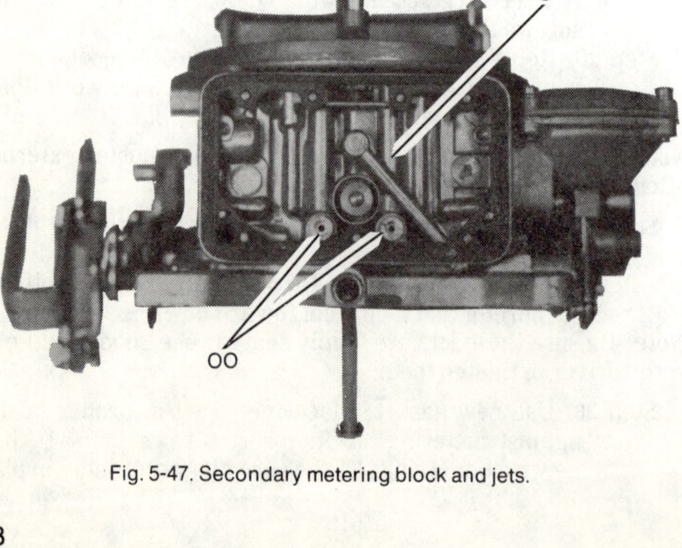

Fig. 5-47. Secondary metering block and jets.

Fig. 5-48. Float-chamber cover.

Step 33. Use new gasket; position primary fuel bowl against metering block J.

Step 34. Use new rubber *O* rings; insert the four screws L, and tighten them evenly while making sure that all gaskets stay in position.

Step 35. Turn carburetor upside down; be sure lever Z (5-52) is positioned as shown.

Step 36. Be sure check ball (5-53) at bottom of accelerator pump diaphragm bowl is clean. Insert compression spring in bowl, insert diaphragm assembly with button visible. Use new gasket; replace accelerator pump

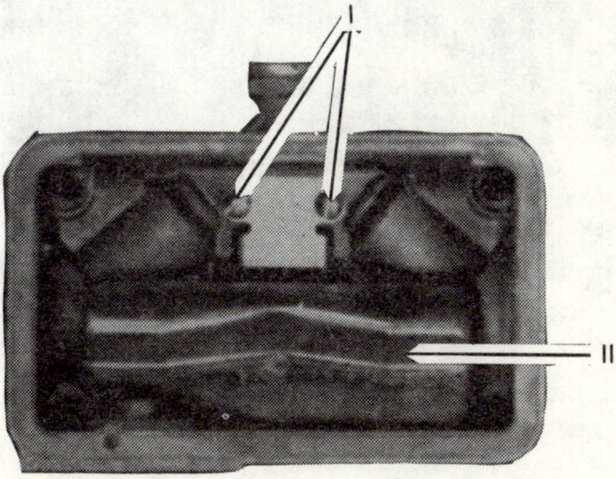

Fig. 5-49. Primary fuel bowl.

Fig. 5-50. Primary metering block, power valve, and jets.

housing AA with lever Z resting on screw head Y (5-52). Tighten screws.

Step 37. Turn carburetor right-side-up. Drop new needle valves H1 and HH1 into wells at E. Place new gaskets in position, and insert adjusting seats H2 and HH2; turn

Fig. 5-51. Float chamber cover and metering block (installed).

Fig. 5-52. Accelerator-pump external linkage.

each seat clockwise until top of needle valve is approximately 1/32 in. below top of nut.

Step 38. Use new gaskets; insert slotted screws E (5-54); tighten loosely with fingers.

Step 39. Install carburetor on car; attach all lines and linkages; but do not install air cleaner on carburetor.

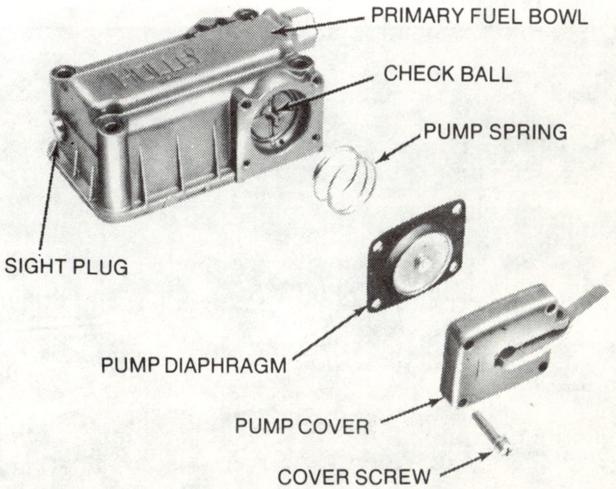

Fig. 5-53. Accelerator-pump assembly sequence.

Fig. 5-54. Float-chamber covers and needles and seats.

Step 40. Park car on level ground; leave engine running.

Step 41. Remove the two screws AA (5-54)—these are called "sight plugs."

Step 42. Adjust fuel level to the bottom of each hole by turning adjusting nuts BB (5-54) clockwise to lower fuel level; counterclockwise to raise fuel level. (Raising fuel level tends to enrich fuel mixture.) When fuel level is adjusted, tighten slotted screws E to lock nut BB. Replace sight plugs AA. Tighten sight plugs and slotted screws firmly.

Fig. 5-55. Idle-speed and mixture-control screws.

Step 43. Adjust idle by turning screw CC (5-55).

Step 44. Adjust idle mixture with idle-mixture adjustment screws DD (5-55); these screws are on each side of primary metering block. Turn clockwise until engine idles rough, then counterclockwise until engine idles smoothly. Adjust one side and then the other until smoothest idle is achieved.

Replace air cleaner to carburetor. Be sure element is clean.

HOLLEY "SPREAD BORE" (Model 4165)

Please refer to Chapter 1 for a brief description and discussion of carburetor cleaners and the solvent container needed to make up the bath to be used for this task. Tools are listed in Table 5-1.

Preliminary Instructions

Remove air cleaner.

Remove all dirt and grime from exterior of carburetor while carburetor is still mounted on engine; use kerosene and a paintbrush if necessary, but **do not use gasoline.**.

Remove carburetor from engine.

For information on carburetor accessories, refer to Chapter 2. If you are not sure you can remember where and how the lines and linkages connect, make notes and sketches in the margins of this book as you remove the carburetor. Figure 5-56, an exploded view, is included to help you identify the various parts of your carburetor.

When carburetor is removed from engine, follow tuneup steps listed below:

To Disassemble

Step 1. Remove small cotter pin A (5-57); detach rod B.

Note: Crossover feed tube C (5-57) supplies fuel from inlet on primary bowl to secondary bowl. This tube is sealed at points D and D1 by Neoprene *O*-rings.

Step 2. Remove large nut E (5-58); remove and discard nut gasket; remove and discard filter from well beneath nut. Remove spring from this well. Place nut and spring in bath.

Step 3. Use broad-blade screwdriver; remove the four screws G (5-58); discard the rubber gaskets from beneath screw heads.

Step 4. Remove primary bowl H (5-58) from main body of carburetor. It may be stuck—tap on bowl with screwdriver handle to loosen it if necessary.

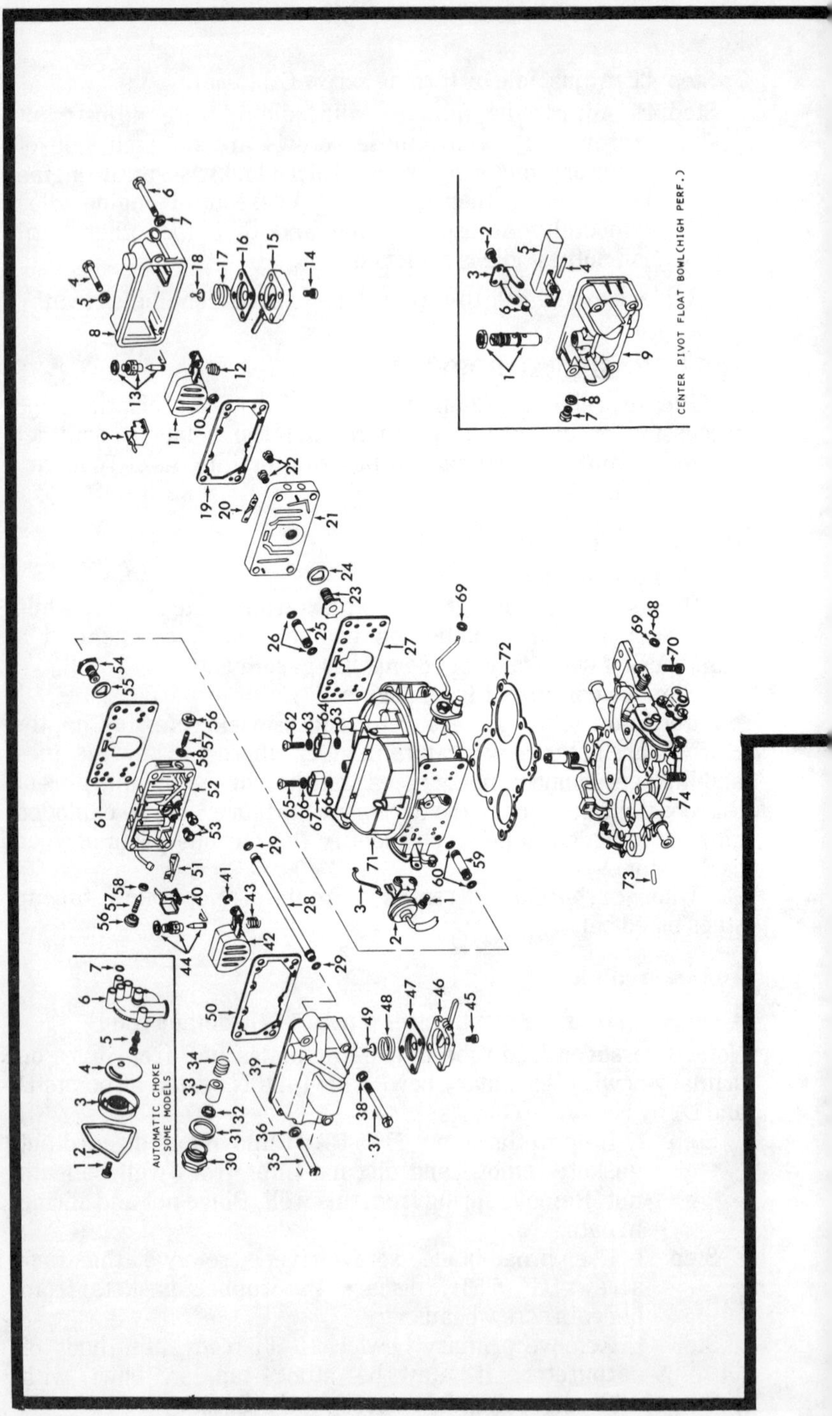
AUTOMATIC CHOKE
SOME MODELS
CENTER PIVOT FLOAT BOWL (HIGH PERF.)

REF. NO.		REF. NO.	
1.	SCREW & LOCKWASHER (2)— CHOKE DIAPHRAGM ASSY.	38.	GASKET (2)— FUEL BOWL SCREW
2.	CHOKE DIAPHRAGM ASSY.	39.	FUEL BOWL ASSY.— PRI.
3.	LINK— CHOKE DIAPHRAGM	40.	BAFFLE PLATE— PRI.
4.	SCREW (SHORT)(2)— SEC. FUEL BOWL	41.	RETAINER— FLOAT
5.	GASKET (2)— FUEL BOWL SCREW	42.	FLOAT & LEVER ASSY.— PRI.
6.	SCREW (LONG)(2)— SEC. FUEL BOWL	43.	SPRING— PRI. FLOAT
7.	GASKET (2)— FUEL BOWL SCREW	44.	NEEDLE, SEAT & GASKET ASSY. PRI.
8.	FUEL BOWL ASSY.— SEC.	45.	SCREW & LOCKWASHER (4)— PRI. PUMP COVER
9.	BAFFLE PLATE— SEC.	46.	COVER— PRI. PUMP
10.	RETAINER— FLOAT	47.	PUMP DIAPHRAGM ASSY.— PRI. (SMALL)
11.	FLOAT & LEVER ASSY.— SEC.	48.	SPRING— PRI. DIAPHRAGM RETURN
12.	SPRING— SEC. FLOAT	49.	VALVE— PRI. PUMP CHECK
13.	NEEDLE, SEAT & GASKET ASSY. SEC.	50.	GASKET— PRI. FUEL BOWL
14.	SCREW & LOCKWASHER (4)— SEC. PUMP COVER	51.	BAFFLE— PRI. METERING BODY VENT
15.	COVER— SEC. PUMP	52.	METERING BODY ASSY.— PRI.
16.	PUMP DIAPHRAGM ASSY.— SEC. (LARGE)	53.	JET (2)— PRI. MAIN
17.	SPRING— SEC. DIAPHRAGM RETURN	54.	VALVE— PRI. POWER
18.	VALVE— SEC. PUMP CHECK	55.	GASKET— PRI. POWER VALVE
19.	GASKET— SEC. FUEL BOWL	56.	CAP (2)— IDLE LIMITER
20.	BAFFLE— SEC. METERING BODY VENT	57.	NEEDLE (2)— IDLE ADJUSTING
21.	METERING BODY ASSY.— SEC.	58.	SEAL (2)— IDLE ADJ. NEEDLE
22.	JETS— SEC. MAIN	59.	TUBE— PRI. PUMP PASSAGE
23.	VALVE— SEC. POWER	60.	O-RING (2)— PASSAGE TUBE
24.	GASKET— SEC. POWER VALVE	61.	GASKET— PRI. METERING BODY.
25.	TUBE— SEC. PUMP PASSAGE	62.	SCREW— SEC. PUMP DISC. NOZZLE
26.	O-RING (2)— PASSAGE TUBE	63.	GASKET (2)— SCREW
27.	GASKET— SEC. METERING BODY	64.	NOZZLE— SEC. PUMP DISCHARGE (LARGE DUMP HOLES)
28.	TUBE— FUEL LINE	65.	SCREW— PRI. PUMP DISC. NOZZLE
29.	O-RING (2)— FUEL LINE TUBE	66.	GASKET (2)— SCREW
30.	FITTING— FUEL INLET	67.	NOZZLE— PRI. PUMP DISCHARGE (SMALL DUMP HOLES)
31.	GASKET— INLET FITTING	68.	COTTER PIN— SEC. CONNECTING ROD
32.	GASKET— FUEL FILTER	69.	WASHER (2)— SEC. CONNECTING ROD
33.	FILTER— FUEL	70.	SCREW & LOCKWASHER (8)— THROTTLE BODY
34.	SPRING— FUEL FILTER	71.	MAIN BODY ASSY.
35.	SCREW (SHORT)(2)— PRI. FUEL BOWL	72.	GASKET— THROTTLE BODY
36.	GASKET (2)— FUEL BOWL SCREW	73.	PLUG— VACUUM TUBE
37.	SCREW (LONG)(2)— PRI. FUEL BOWL	74.	THROTTLE BODY ASSY.

Fig. 5-56. Exploded view of Holley "Spread Bore" four-barrel carburetor, Model 4165.

Note: The crossover feed tube will disconnect at this time; remove it completely from both bowls, and remove and discard the *O*-rings from either end of it.

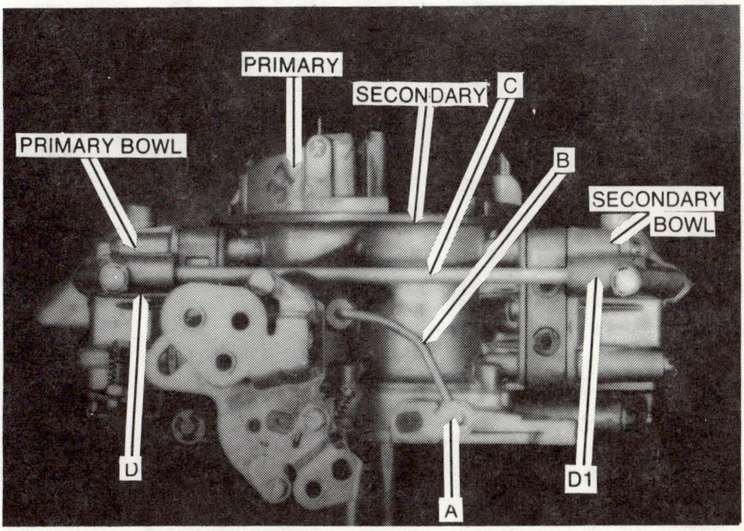

Fig. 5-57. Cross-feed tube and linkage.

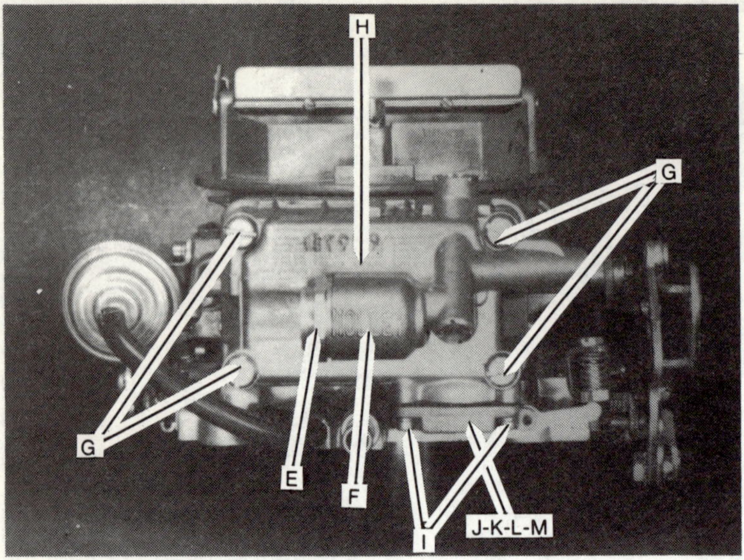

Fig. 5-58. Fuel filter, primary bowl, bowl screws, and related parts.

Step 5. Remove the four screws I (5-58).

Step 6. Remove accelerator pump housing J (5-58); from well beneath housing, remove and discard diaphragm K; and from beneath diaphragm, remove coil spring L and place it in bath. Finally, from bottom of well, pull out mushroom-shaped check valve M and discard it.

Note: This valve protrudes through bowl wall, thus some effort is required to free it.

Step 7. Remove *E*-clip N (5-59) from float-hinge shaft.

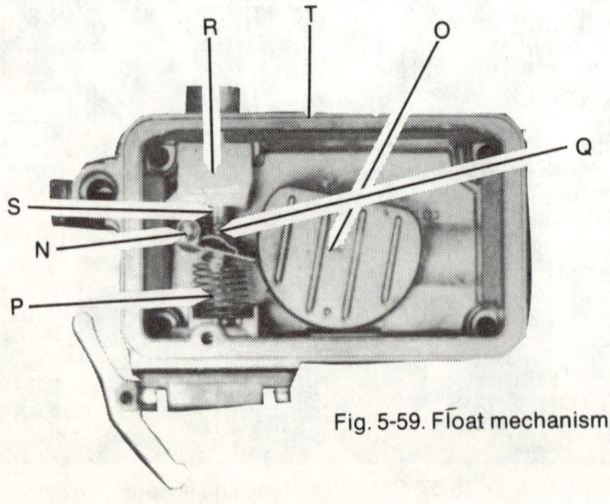

Fig. 5-59. Float mechanism.

Fig. 5-60. Metering block, outboard side.

Step 8. Remove float O (5-59); place it in bath.
Step 9. Remove coil spring P (5-59); place it in bath.
Step 10. Remove and discard needle valve Q (5-59).
Step 11. Remove plastic splash baffle R (5-59).
Step 12. Remove and discard needle-valve seat S (5-59).
Step 13. Place primary bowl T (5-59) in bath.
Step 14. Remove primary metering block U (5-60) from main body of carburetor. It may be stuck—tap on block with screwdriver handle to loosen it if necessary.
Step 15. Remove jets V (5-60) *only* if they evidence damage or if you plan to replace them with larger or smaller jets. (New jets are not included in repair parts kit.)
Step 16. Remove rubber *O*-ring from brass connecting tube W (5-61).
Step 17. Remove power valve X (5-61); use adjustable wrench—turn counterclockwise. Discard valve.

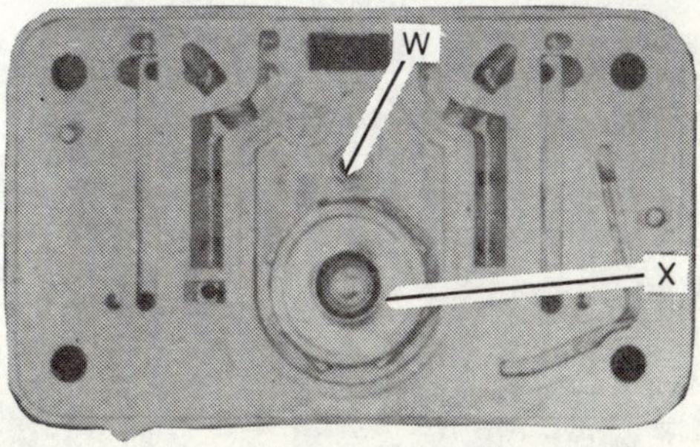

Fig. 5-61. Metering block, inboard side.

Fig. 5-62. Choke pull-off and linkage.

Step 18. Place metering block in bath.

Step 19. The primary and secondary bowls and metering blocks are essentially mirror images of one another; therefore, to disassemble the secondary side of your carburetor, repeat steps 3 through 18, substituting the word "secondary" for "primary" wherever it appears. But *do not get the parts from the two mixed up—keep them separate at all times, especially in the bath.*

Step 20. Remove vacuum hose Y (5-62).

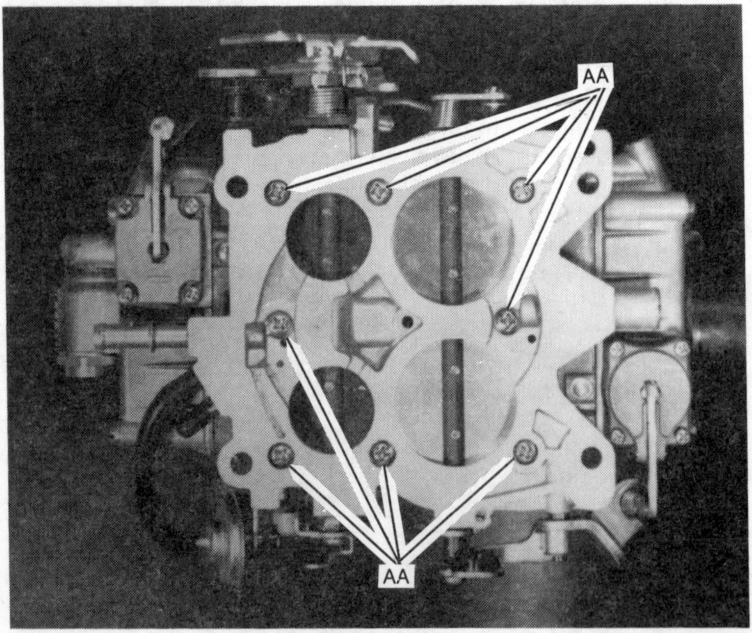

Fig. 5-63. Base-to-body screws.

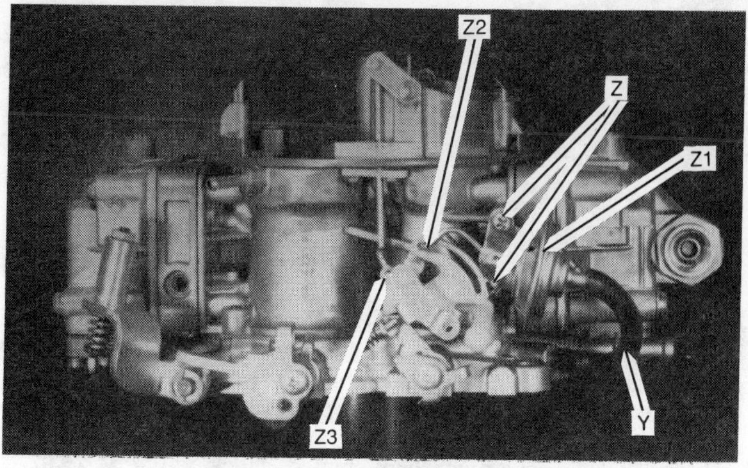

Fig. 5-64. Choke pull-off and linkage.

Step 21. Remove the two screws Z (5-62).

Step 22. Remove choke pull-off Z1 (5-62) and, at the same time, disconnect choke pull-off linkage at Z2. Do *not* place choke pull-off in bath.

Step 23. Remove the eight screws AA (5-63); separate base from main body of carburetor; discard gasket; place both carburetor parts in bath.

Note: Allow all parts placed in bath to be cleaned properly in accordance with directions on carburetor-cleaner container before beginning reassembly. Then, remove all parts from cleaner, rinse with water, and allow to dry naturally. Do not use compressed air.

To Reassemble

Step 24. Use new gasket. Position main body of carburetor on base and assemble the two parts together with the eight screws AA (5-63).

Note: Choke rod must be attached at Z3 (5-63) before screws are tightened.

Step 25. Install choke pull-off assembly Z1 (5-64); reattach linkage at Z2, then insert and tighten the two screws Z.

Step 26. Install a new vacuum hose Y (5-64).

Note: This hose is not included in repair parts kit; ask for 5 in. of $^3/_{16}$ in. vacuum hose.

Step 27. Install new power valve X (5-65)—use new gasket; tighten firmly with adjustable wrench.

Step 28. Install a new *O*-ring over brass connecting tube W (5-65).

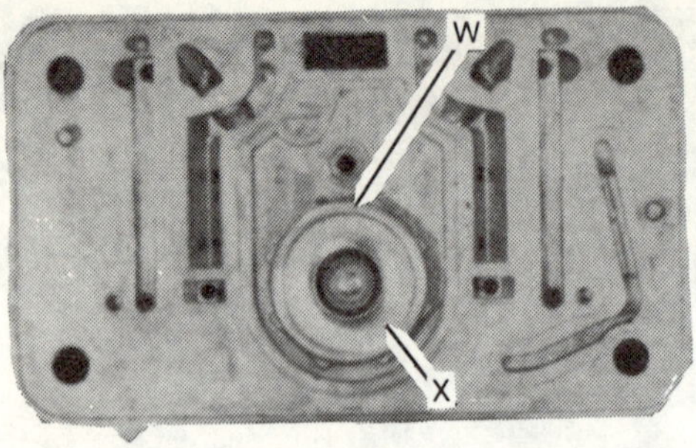

Fig. 5-65. Metering block, inboard side.

Step 29. Use new gasket, position metering block U (5-66) on primary end of the main carburetor body so that the two jets V remain visible.

Step 30. Use new mushroom-shaped check valve M (5-67); lubricate it with oil (or water) and insert it into hole in bottom of accelerator pump well.

Step 31. Position coil spring L (5-67) into well and over valve M.

Step 32. Place diaphragm K (5-67) into well over spring.

Note: Be sure the round center button on the diaphragm is turned outward (not in toward spring).

Step 33. Position accelerator pump housing J (5-67) over diaphragm and secure it in place with the four screws I. Tighten screws firmly.

Step 34. Install new needle-valve seat S (5-67). Tighten firmly.

Step 35. Install plastic baffle R (5-67).

Step 36. Install new needle valve Q (5-67) into seat S.

Step 37. Install spring P (5-67) into seat S.

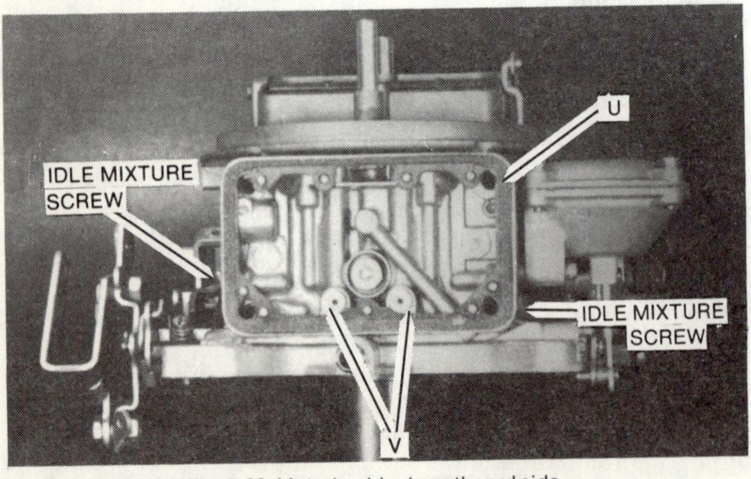

Fig. 5-66. Metering block, outboard side.

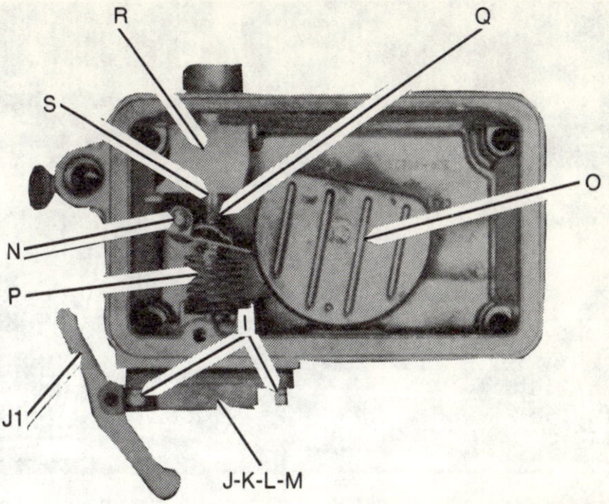

Fig. 5-67. Float, accelerator pump housing, and related parts.

Step 38. Replace the *E*-clip N (5-67) on end of float-hinge shaft.

Step 39. Set float level; information for setting float level is provided on data sheet contained in repair parts kit.

Step 40. Use new gasket. Position bowl on metering block; be sure lever J1 is properly positioned below accelerator adjusting screw ZZ, as shown in 5-68.

Step 41. Install the screws G (5-68)—be sure to use new gaskets under screw heads, and remember that the

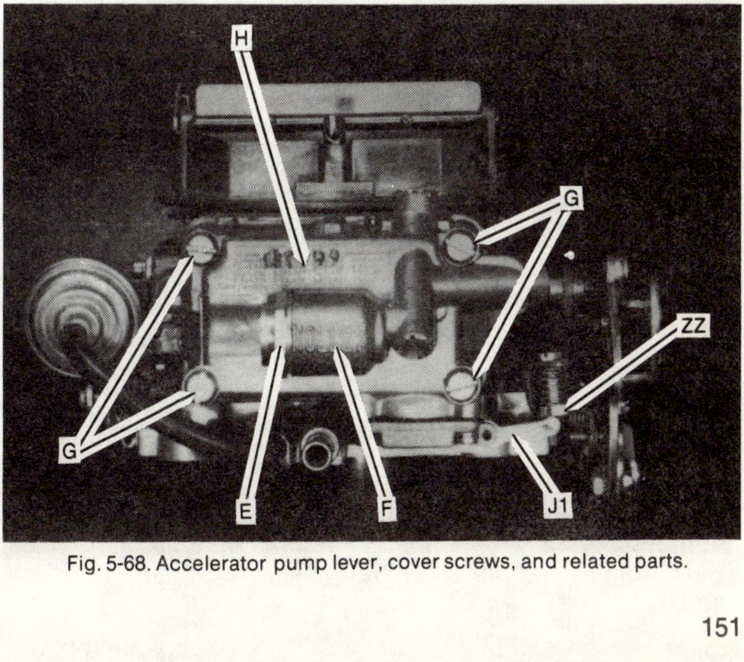

Fig. 5-68. Accelerator pump lever, cover screws, and related parts.

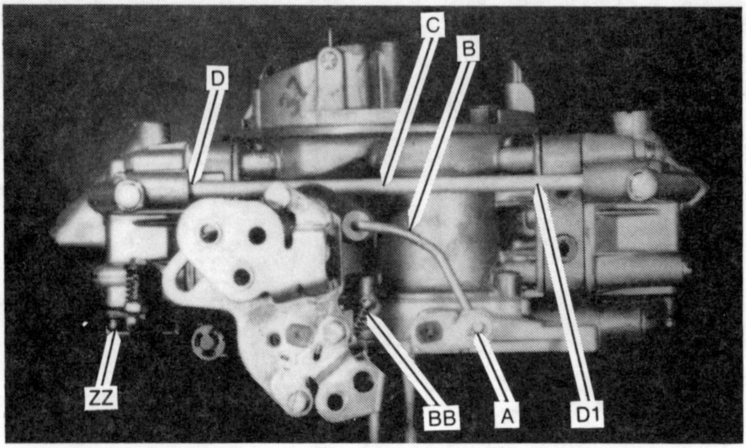

Fig. 5-69. Cross-feed tube, linkage, and accelerator-pump adjusting screw.

longer screws go in the lower holes. Tighten screws very firmly.

Step 42. Place new *O*-rings on tip ends of crossover feed tube C (5-69) and insert one end of tube into primary bowl at point D1. Then, remember to insert other end into secondary bowl as it is assembled to carburetor.

Step 43. Repeat steps 27 through 41, substituting the word "secondary" for "primary" wherever it appears.

Step 44. Attach linkage B (5-69) at A; install new cotter pin A.

Step 45. Install coil spring into well F (5-70).

Step 46. Install new filter into well F (5-70) on top of spring.

Note: Insert filter flat end first, with circular opening toward the outside.

Fig. 5-70. Filter assembly.

Fig. 5-71. Idle-speed and mixture-adjustment screws.

Step 47. Use new gasket. Install and tighten nut E (5-70).

Step 48. Mount carburetor on engine. Be sure all lines and linkages are properly attached.

Step 49. Start engine. With engine idling, check visually for signs of gasoline leaks. If leaks appear, turn engine off and tighten all outward screws. If leak persists, disassemble carburetor, check all gaskets and rubber parts—replace as needed—and reassemble, making sure that all screws are tightened firmly. Mount carburetor on engine, and check again for leaks.

Step 50. Allow engine to warm up and primary choke valves to fully open.

Step 51. Adjust idle to desired speed by rotating screw BB (5-71).

Step 52. Adjust idle mixture with screws DD (5-71) at either side of primary metering block. Run one screw in until idle begins to become rough, then back it out until smoothest idle is obtained; repeat this procedure with other screw. Alternate back and forth between the two screws until you are satisfied with idle.

Install the air cleaner. Be sure the element is clean.

Chapter 6

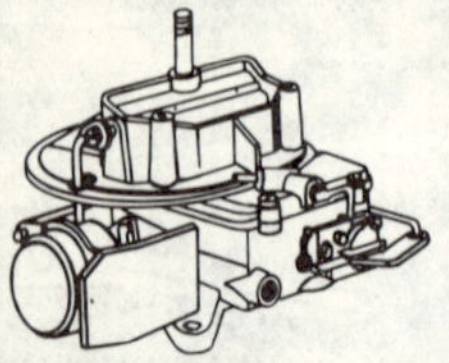

Rochester

Rochester Products, a division of General Motors, produces a wide range of automotive hardware from glovebox locks to emission control devices. Rochester carburetors are routinely fitted to GM cars and trucks and have appeared on other brands as well. The four-barrel Quadrajet is generally recognized as the most sophisticated carburetor offered to the American motoring public.

ROCHESTER ONE-BARRELS (MODELS B, BC, AND BV)

Please refer to Chapter 1 for a brief description and discussion of carburetor cleaners and the solvent container to make up the bath to be used for this task.

Only a few tools are required for carburetor tuneup, as shown in Table 6-1.

Table 6-1. Tools Required for Rochester Carburetors Overhaul.

Screwdriver with a broad blade ($^5/_{16}$ in. or larger).
Screwdriver with a medium Phillips head.
Adjustable wrench with at least a 1 in. span.
Needle-nosed pliers.
Open-end, box-end, or socket wrench of the proper size to fit the nuts or bolts that hold your carburetor to the manifold.
Note: Most air cleaners can be removed without tools or with only a pair of pliers.

Preliminary Instructions

Remove air cleaner.

Brush dirt and grime from carburetor while it is mounted on engine; use kerosene and stiff paintbrush if necessary, but **do not use gasoline.**

Remove carburetor from engine.

For information on carburetor accessories, refer to Chapter 2. If you're not sure you can remember where and how the lines and linkages connect, make notes and sketches in the margin of this book as you remove the carburetor. Figure 6-1, an exploded

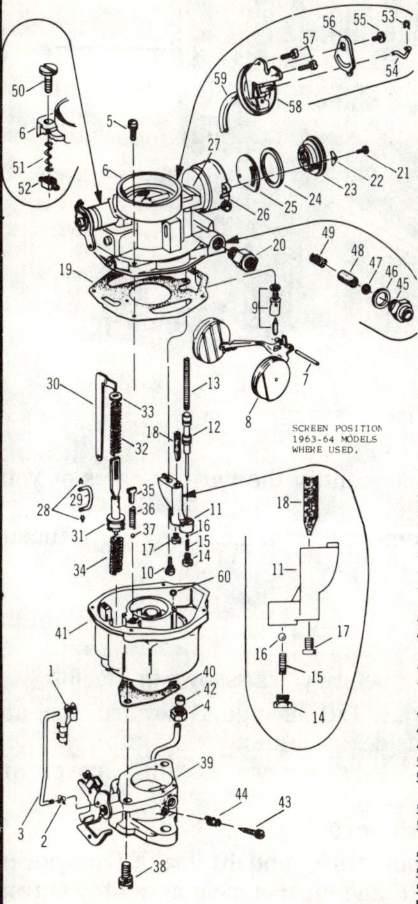

REF.
NO.

1. RETAINER-CHOKE ROD (UPPER)
2. RETAINER-CHOKE ROD (LOWER)
3. ROD-CHOKE
4. NUT-CHOKE SUCTION TUBE
5. SCREW & WASHER-BOWL COVER
6. BOWL COVER ASSEMBLY
7. PIN-FLOAT LEVER HINGE
8. FLOAT & LEVER ASSEMBLY
9. NEEDLE, SEAT & GASKET ASSY.
10. SCREW & WASHER-MAIN WELL SUPPORT
11. SUPPORT ASSY.-MAIN WELL
12. PISTON-POWER
13. SPRING-POWER PISTON
14. PLUG-POWER VALVE
15. SPRING-POWER VALVE
16. BALL-POWER VALVE
17. JET-MAIN METERING
18. SCREEN-MAIN WELL SUPPORT
19. GASKET-BOWL COVER
20. FITTING-FUEL INLET
21. SCREW-STAT COVER
22. RETAINER-STAT COVER
23. STAT COVER & SPRING ASSY.
24. GASKET-STAT COVER
25. PLATE-CHOKE BAFFLE
26. CHOKE HOUSING
27. GASKET CHOKE HOUSING (NOT SHOWN)
28. RETAINER-PUMP LINK
29. LINK-PUMP
30. ROD-PUMP
31. PUMP PLUNGER
32. SPRING-PUMP PLUNGER
33. WASHER-PUMP SPRING
34. SPRING-PUMP RETURN
35. GUIDE-PUMP DISCHARGE BALL
36. SPRING-PUMP DISCHARGE BALL
37. BALL-PUMP DISCHARGE
38. SCREW & WASHER-THROTTLE BODY TO BOWL
39. THROTTLE BODY ASSEMBLY
40. GASKET-BODY FLANGE
41. BOWL-FLOAT
42. PACKING-CHOKE SUCTION TUBE
43. NEEDLE-IDLE ADJUSTING
44. SPRING-IDLE ADJUSTING NEEDLE
45. FITTING-FUEL INLET
46. GASKET-FUEL INLET FITTING
47. GASKET-FUEL FILTER
48. FILTER-FUEL INLET
49. SPRING-FUEL INLET FILTER
50. VALVE-IDLE VENT (BV)
51. SPRING-IDLE VENT VALVE (BV)
52. NUT-IDLE VENT VALVE (BV)
53. RETAINER-VACUUM CONTROL ROD (BV)
54. ROD-VACUUM CONTROL (BV)
55. SCREW-STAT ROD LEVER (BV)
56. LEVER-STAT ROD (BV)
57. SCREW-VACUUM CONTROL ATTACHING (BV)
58. VACUUM BREAK CONTROL (BV)
59. HOSE-VACUUM BREAK CONTROL (BV)
60. O-RING-VACUUM CHANNEL TUBE (PARTIAL-STARTING 1965 MODELS)

Fig. 6-1. Exploded view of Rochester one-barrel carburetor, Models B, BC, and BV.

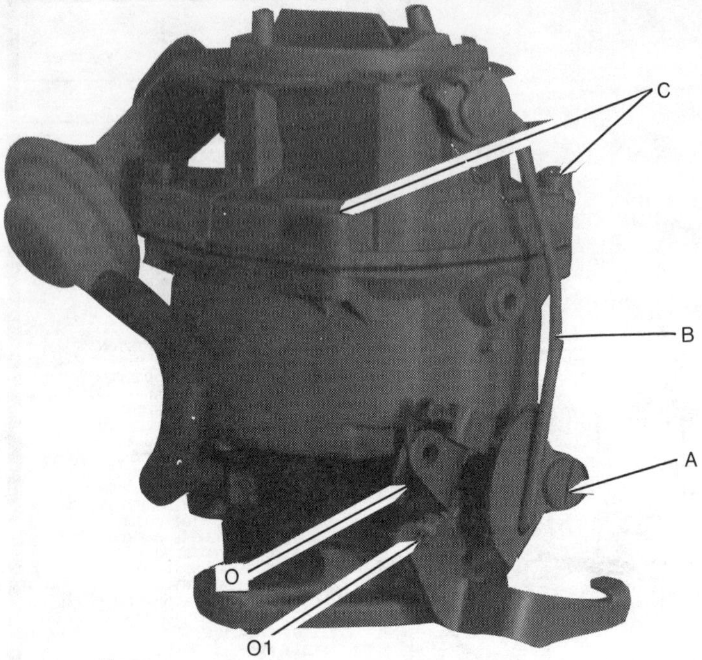

Fig. 6-2. Linkage and cover screws.

view, is included to help you identify the various parts of your carburetor.

When carburetor is removed from engine, follow tuneup steps listed below.

To Disassemble

Step 1. Place carburetor before you as shown in Fig. 6-2.
Step 2. Remove screw A. Lift linkage B toward you and remove it. Lay it aside.
Step 3. Remove four cover screws C. Lift cover off carburetor bowl.
Step 4. Place cover as shown in Fig. 6-3.
Step 5. Remove hinge pin D (6-3), and lift float E from cover.
Step 6. Remove screw F and lift metering assembly G from cover.
Step 7. Remove screw H (6-3). *Carefully* remove spring I and check ball J from well J1. Place spring and ball in bath.
Step 8. Remove vacuum piston K (6-3) and spring L from cover. Place piston in bath.
Step 9. Remove and discard the bowl cover gasket.

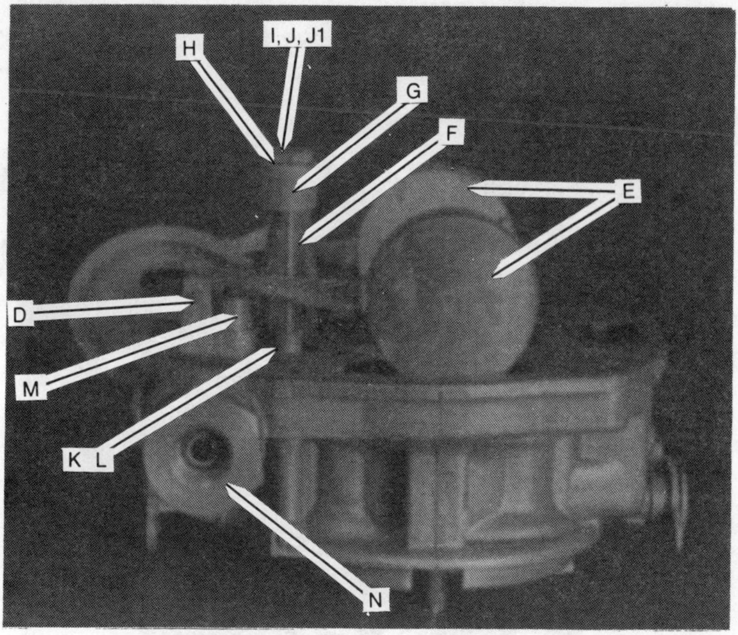

Fig. 6-3. Bowl-cover assembly.

Step 10. Remove needle-and-seat assembly M (6-3). Discard this assembly.

Step 11. Remove nut N (6-3). Remove and discard filter from inside housing beneath this nut.

Note: This filter is not included in carburetor parts kit, but should be replaced with any tuneup or overhaul.

Step 12. Place bowl cover in bath.

Step 13. Continue teardown by removing clip O1 and linkage O (6-2) from carburetor bowl.

Step 14. Lift accelerator plunger arm P (6-4).

Step 15. Remove plunger Q and spring Q1 (6-4) from arm R. Discard plunger; place spring in bath.

Step 16. Now lift spring (6-4) from pump well O2.

Step 17. Remove retainer T (6-4) by pulling upward with pliers.

Note: This retainer may be quite stubborn, so keep pulling until it releases.

Step 18. From under retainer T, remove spring U and check ball V (6-4).

Step 19. Turn carburetor bowl upside down.

Step 20. Remove the two screws W (6-4) that hold carburetor base to the bowl.

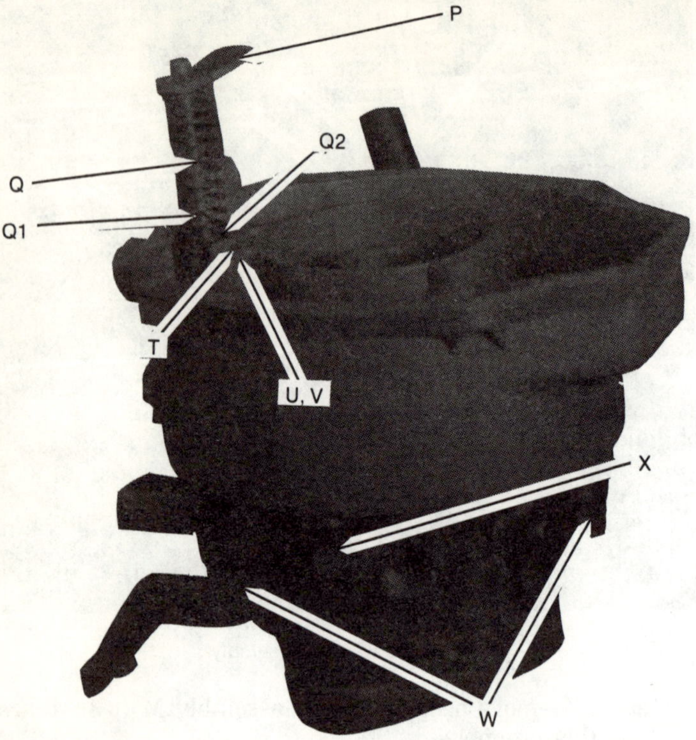

Fig. 6-4. Accelerator pump, base-to-bowl screws, and idle-mixture screw.

Step 21. Remove and discard gasket from between base and bowl. Place bowl in bath.

Step 22. Remove idle-mixture adjustment screw and spring assembly X (6-4) from base. Place assembly in bath.

Step 23. Place carburetor base in bath.

Note: You have completed the disassembly. Allow all parts placed in bath to be properly cleaned in accordance with directions on carburetor-cleaner container before beginning reassembly. The remove all parts from cleaner, rinse with water, and allow to dry naturally. Do not use compressed air.

To Reassemble

Step 24. Place carburetor base before you as shown in Fig. 6-4.

Step 25. Insert idle-mixture adjustment screw and spring assembly X (6-5) into base. Tighten screw with *fingers only* until needle is seated, then back it out one complete turn.

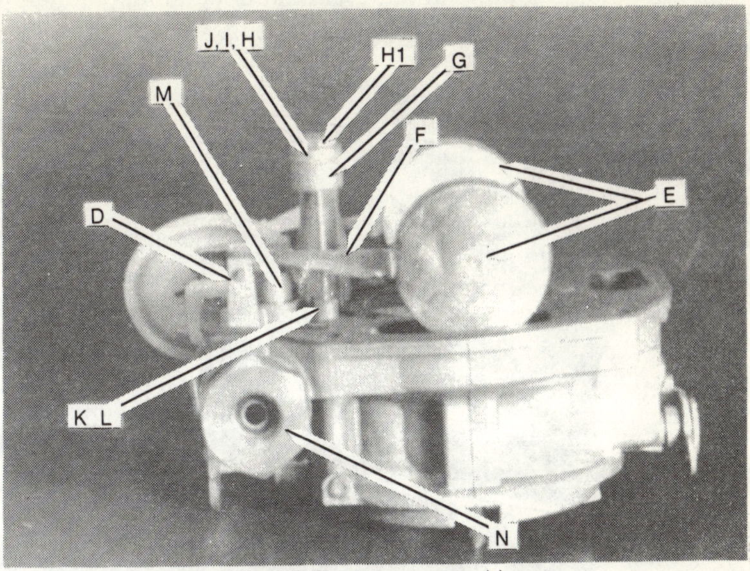

Fig. 6-5. Bowl-cover assembly.

Step 26. Place new gasket on base end of bowl; place base on bowl; insert the two screws W (6-4) that hold base to bowl; tighten screws firmly.

Step 27. From your kit of new parts, select the proper check ball and drop it into hole Y (6-4). Replace spring U and retainer T. (You may need to refer back to steps 17 and 18.)

Step 28. Drop spring Q2 into accelerator pump well (6-4).

Step 29. Install new accelerator plunger Q and spring Q1 on arm P disassembled in step 15.

Step 30. Push accelerator plunger and arm into well (6-4). Replace linkage O (6-6) with clip O1.

Step 31. Place bowl cover assembly before you as in 6-5.

Step 32. Install new needle-and-seat assembly M (6-5) over new gasket. Tighten firmly with screwdriver.

Step 33. Install new bowl-cover gasket.

Step 34. Install new spring L and vacuum piston K (6-5).

Step 35. Install new check ball J and spring I (6-6) into well J1—secure firmly with screw H.

Step 36. Place entire metering assembly G down over vacuum piston and into proper position. Replace screw F (6-5); tighten firmly.

Step 37. Double-check and retighten retainer screw H (6-5), and tighten metering jet H1 adjacent to it.

Step 38. Inspect floats to be sure they contain no liquid and are not collapsed on the sides. Install float assembly E (6-5) and hinge pin D removed in step 5).

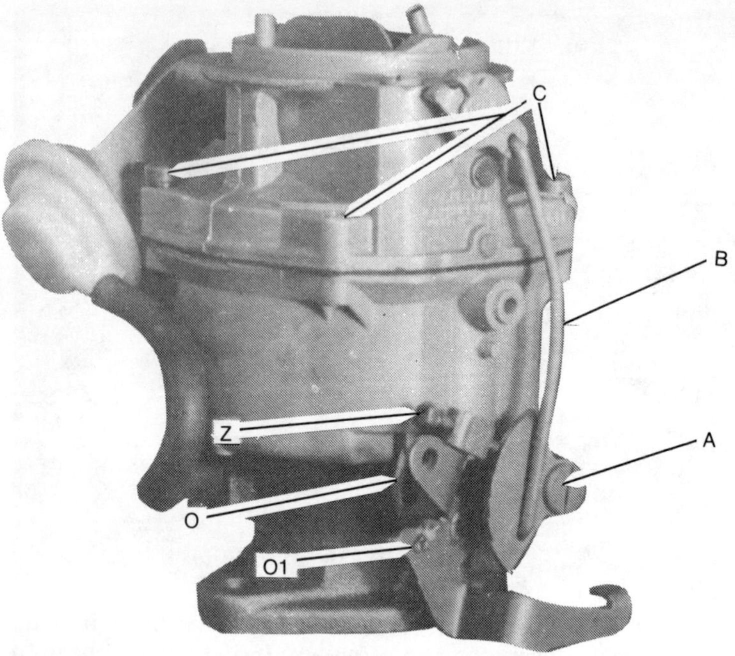

Fig. 6-6. Linkage and bowl-cover screws.

Step 39. Make float level adjustment in accordance with instructions provided in carburetor parts kit. *Proper float level adjustment is very important.*

Step 40. Replace cover on bowl assembly. Tighten all screws C (6-6) firmly.

Step 41. Replace linkage B and screw A (6-6).

Step 42. Install new filter element and replace nut N (6-5) over new gasket.

Step 43. Install carburetor on engine. Remake all connections as noted in preliminary instructions.

Step 44. Start engine, allow to warm until choke opens. Make idle-speed adjustment by turning screw Z (6-6) until engine idles at preferred speed.

Step 45. To insure proper idle mixture, turn screw X (6-4) clockwise until engine begins to run rough, then counterclockwise until engine runs smoothly.

Replace air cleaner. Be sure element is clean.

ROCHESTER MONOJET (MODEL MV)

Please refer to Chapter 1 for a brief description and discussion of carburetor cleaners and the container needed to make up the bath to be used for this task.

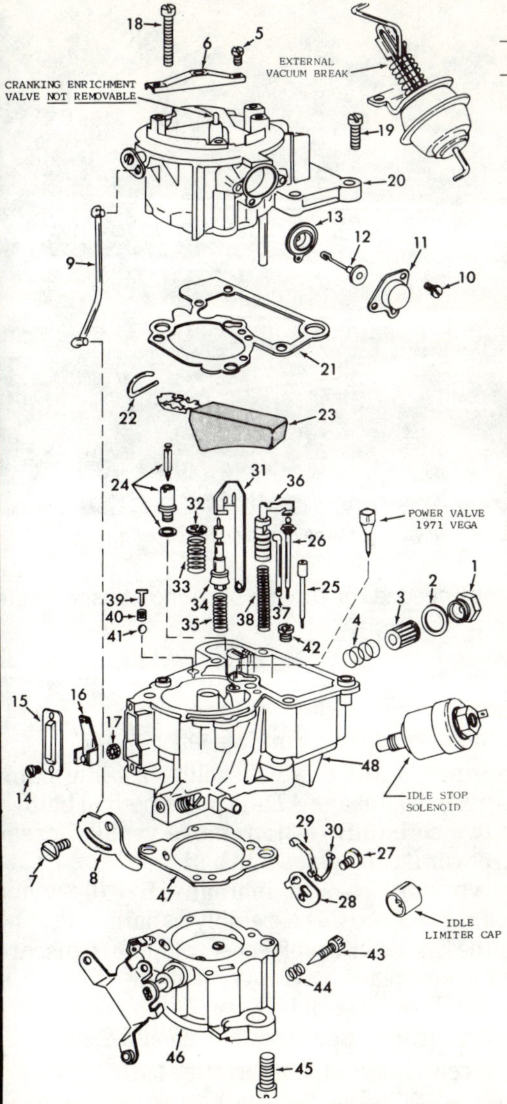

NOMENCLATURE

REF. NO.	
1.	NUT - FUEL INLET FILTER
2.	GASKET - FILTER NUT
3.	FILTER - FUEL
4.	SPRING - FUEL FILTER
5.	SCREW(2) -AIR CLEANER STUD BRACKET
6.	BRACKET - AIR CLEANER STUD
7.	SCREW - FAST IDLE CAM
8.	CAM - FAST IDLE
9.	ROD - CHOKE
10.	SCREW(2) - DIAPH. COVER
11.	COVER - DIAPHRAGM
12.	LINK ASSY. - VACUUM BREAK DIAPHRAGM
13.	DIAPHRAGM - VACUUM BREAK
14.	SCREW (2) - IDLE COMPENSATOR COVER
15.	COVER - IDLE COMPENSATOR
16.	IDLE COMPENSATOR ASSY.
17.	GASKET - IDLE COMPENSATOR
18.	SCREW LONG(3) - BOWL COVER
19.	SCREW SHORT(3)-BOWL COVER
20.	BOWL COVER
21.	GASKET - BOWL COVER
22.	HINGE PIN - FLOAT
23.	FLOAT ASSEMBLY
24.	NEEDLE, SEAT & GASKET ASSY.
25.	IDLE TUBE ASSEMBLY
26.	METERING ROD & SPRING
27.	SCREW - PUMP & POWER LINK LEVER
28.	LEVER - PUMP & POWER LINK
29.	LINK - PUMP LEVER
30.	LINK - POWER PISTON ROD
31.	LEVER - PUMP OPERATING
32.	RETAINER - PUMP SPRING
33.	SPRING - PUMP
34.	PUMP CUP & STEM ASSEMBLY
35.	SPRING - PUMP RETURN
36.	POWER PISTON ASSEMBLY
37.	ROD - POWER PISTON
38.	SPRING - POWER PISTON
39.	GUIDE - PUMP DISCH. SPRINC
40.	SPRING - PUMP DISCHARGE
41.	BALL - PUMP DISCHARGE
42.	JET - MAIN METERING
43.	NEEDLE - IDLE ADJUSTING
44.	SPRING - IDLE NEEDLE
45.	SCREW (2) - THROTTLE BODY
46.	THROTTLE BODY ASSEMBLY
47.	GASKET - THROTTLE BODY
48.	FLOAT BOWL ASSEMBLY

Fig. 6-7. Exploded view of a Rochester Monojet carburetor, Model MV.

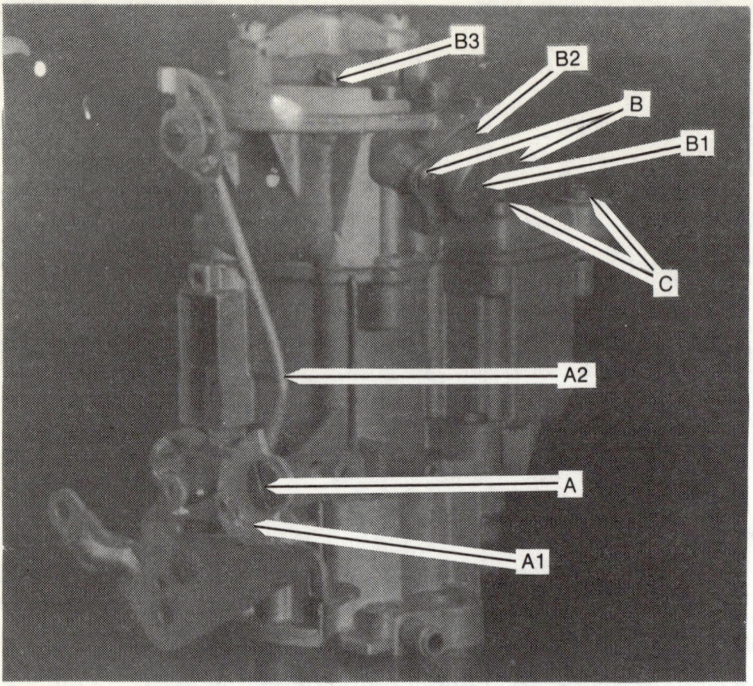

Fig. 6-8. Linkage, vacuum break, and cover screws.

Only a few tools are needed for tuneup of Monojet (see Table 6-1).

To Disassemble

Step 1. Place carburetor as shown in Fig. 6-8.

Step 2. Remove screw A (6-8), which holds fast-idle cam; remove cam A1 and linkage A2 and place both in bath.

Step 3. Remove two screws B (6-8); remove vacuum-break diaphragm cover B1 and place it in bath.

Step 4. Remove vacuum-break diaphragm B2 (6-8) and linkage B3. Do *not* place this assembly in bath.

Step 5. Remove the six bowl cover screws C (6-8); discard bowl cover gasket; place bowl cover in bath.

Step 6. Remove idle-tube assembly D (6-9). To do so, turn carburetor bowl over—tube assembly should drop out.

Step 7. Remove screw E (6-10) from throttle shaft F.

Note: Pay careful attention to arrangement of linkage E1 held in place by screw E; then remove linkage and place it in bath—upward spring pressure from accelerator pump G and metering rod assembly H may cause some difficulty in removing linkage, but it must be wrestled free.

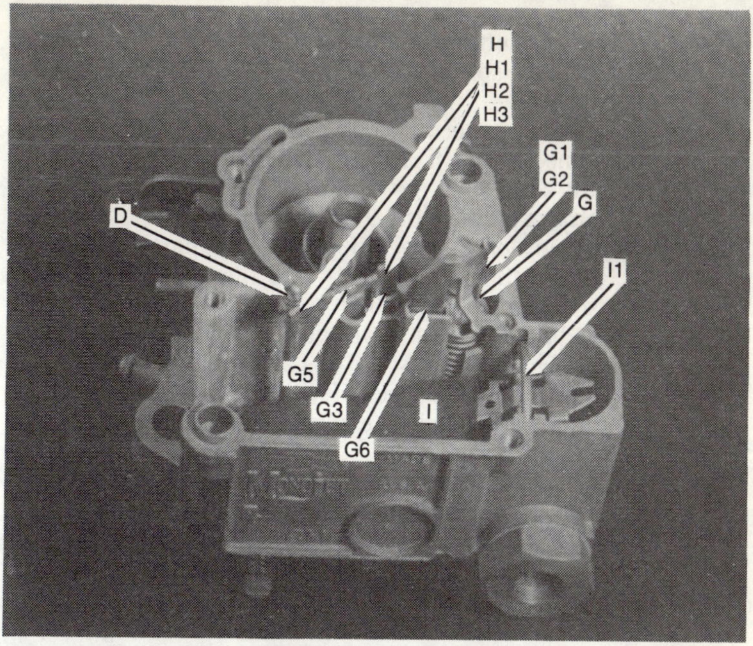

Fig. 6-9. Bowl assembly.

Fig. 6-10. Bowl assembly less float and accelerator pump parts.

Step 8. Lift accelerator pump assembly G (6-9) out of well G1; remove spring G2 from well G3 (6-9). Place spring in bath.

Step 9. Remove metering rod assembly H (6-9)—do not disconnect piston from rod—and place entire assembly in bath.

Step 10. Remove spring H1 (6-9) from well H3 (6-9) and rod H2 from hole alongside well H3. Place both parts in bath.

Step 11. Remove float retainer I1 and float I (6-9); do *not* place float assembly in bath.

Step 12. Use broad-blade screwdriver to remove needle-valve assembly J (6-10); discard this assembly.

Step 13. Tighten metering jet K (6-10), but do not remove it unless it shows obvious evidence of damage or unusual wear.

Step 14. Use needle-nosed pliers to remove check-valve retainer L (6-10). From the well beneath this retainer, remove and discard small coil spring L1 and steel ball 12.

Step 15. Use adjustable wrench to remove large fitting M (6-10); discard filter M1 from well.

Fig. 6-11. Accelerator pump linkage, mixture screw, base-to-bowl screws.

Note: This filter is not included in your carburetor parts kit—ask for it separately.

Step 16. Remove two large screws N (6-11) and separate carburetor bowl from base. Discard gasket. Place bowl in bath.

Step 17. Remove idle-mixture adjustment screw O (6-11) and spring O1 from well O2 in carburetor base. Place both in bath.

Step 18. Place carburetor base in bath.

Diassembly of your carburetor is complete. Allow all parts placed in bath to be properly cleaned in accordance with directions on carburetor cleaner-container before beginning reassembly. The remove all parts from cleaner, rinse with water, and allow to dry naturally. Do not use compressed air.

To Reassemble

Step 19. Install idle-mixture adjustment screw O (6-11) in spring O1 and insert both into well O2; tighten screw until it seats lightly, then back it out 4 turns.

Step 20. Use two screws N (6-11) and a new gasket; assemble carburetor bowl to base.

Note: Always compare new gasket with old to be sure new one has all the holes in the right places.

Step 21. Use new gasket; install a new needle-valve seat J2 (6-12) into well J1. Tighten seat firmly with broad-blade screwdriver.

Step 22. Drop new needle valve J3 (6-12) into seat.

Note: Most needle valves have a resilient tip—make sure this tip goes into seat first.

Step 23. Drop float assembly (float I and retaier I1) into position (6-13).

Step 24. Make float level adjustment. Refer to 6-14 and make adjustment in accordance with instructions contained in your carburetor parts kit. This adjustment is extremely important, so make it with care.

Step 25. Drop new steel ball L2 (6-13) and new coil spring L1 into well L3 (spring must go in last).

Step 26. Place retainer L (6-13) into spring L1, tap retainer lightly into position to secure it in place.

Step 27. Drop coil spring H1 (6-13) into well H3.

Step 28. Drop metering rod linkage H2 into hole alongside well H3.

Step 29. Place metering rod and piston assembly H (6-13) into position. To do so, just raise linkage H2 to fit into slot in piston as piston is placed over spring H1 and pushed into

Fig. 6-12. Bowl-cover assembly, less float and accelerator pump.

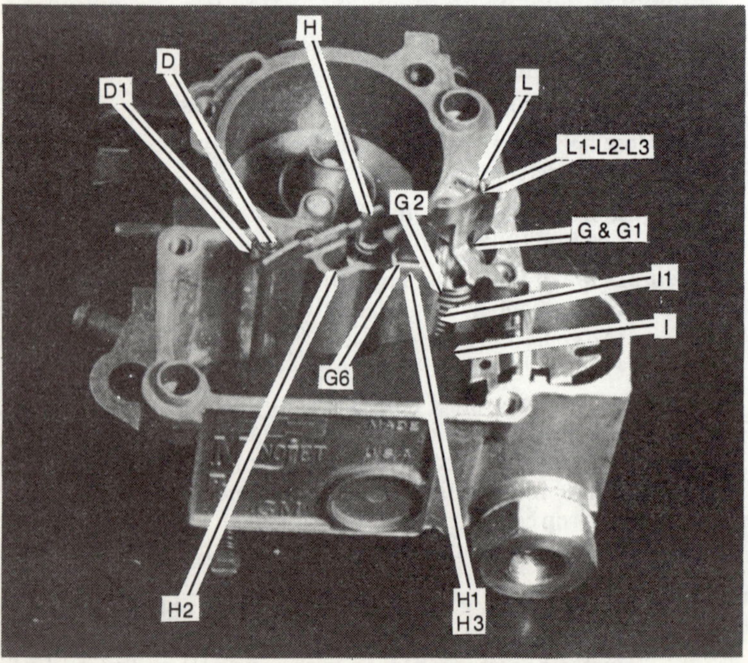

Fig. 6-13. Pump, metering rod, and float.

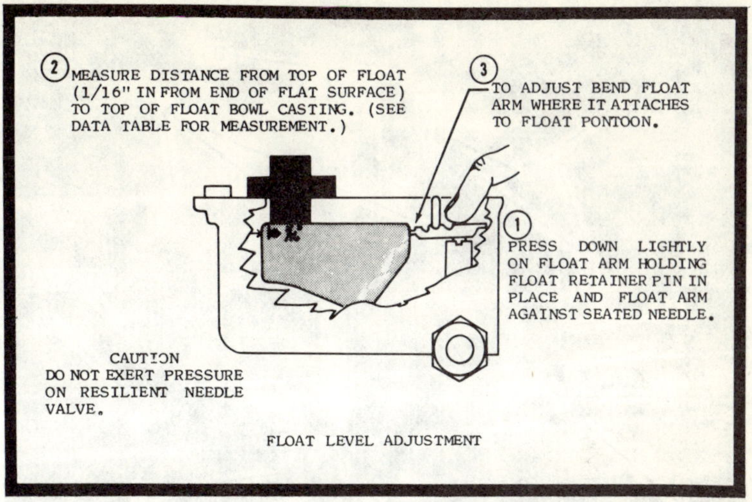

Fig. 6-14. Float-level adjustment.

position. Now, be sure metering rod is aligned with jet K (6-10) in bottom of bowl. If all is in order, push entire assembly downward to be sure it moves freely. Finally, place a matchstick or some such suitable item through hole in bottom of rod to hold assembly in place temporarily.

Step 30. Place coil spring G2 (6-13) into well G1.

Step 31. Remove clip G3 and spring G4 from old accelerator pump assembly G and assemble spring and clip to new plunger.

Step 32. Place new accelerator pump assembly G (6-13) into well G1 over spring G2.

Step 33. Place pump activating arm E2 (6-13) into slot G6 (6-13) and down over plunger; press down on arm to force plunger into well. Be sure this assembly works freely then, while holding assembly down, place another matchstick through hole at lower end of arm to hold assembly in position as you go on to the next steps.

Step 34. Place idle tube assembly D (6-13) into well D1, *point down.*

Note: Blow through tube before you install it to be sure it is not clogged.

Step 35. Before proceeding further, recheck float and metering rod assembly for alignment and freedom of movement. (You may wish to refer back to steps 24 and 29.)

Step 36. Install linkage E1 (6-15) and secure it in place with screw E. Tighten screw firmly.

Fig. 6-15. External components.

Step 37. Place new bowl cover gasket on bowl. Place cover into position and secure it with six screws C (6-15). Tighten screws firmly.

Step 38. Remove old rubber diaphragm B2 (6-15) from shaft B3 and install new one.

Step 39. Place new diaphragm assembly and linkage B3 into position; place cover B1 into position; secure cover in place with two screws B. Tighten screws firmly.

Step 40. Install fast-idle cam A1 (6-15) and linkage A2; secure both in place with screw A. Tighten screw firmly.

Step 41. Place new filter M1 (6-15) into well M2 and secure in place with fitting M. Tighten fitting firmly.

Step 42. Install carburetor on engine.

Step 43. Start engine. Allow engine to warm up until choke valve opens fully.

Step 44. Adjust idle speed by turning screw P (6-15).

Step 45. Adjust idle mixture with screw O (6-15). Turn screw clockwise until engine begins to falter, then back it out counterclockwise until engine runs smoothly.

Replace the air cleaner. Be sure element is clean.

ROCHESTER TWO-BARRELS (MODELS 2G, 2GC, & 2GV)

Please refer to Chapter 1 for a brief description and discussion of carburetor cleaners and the container needed to

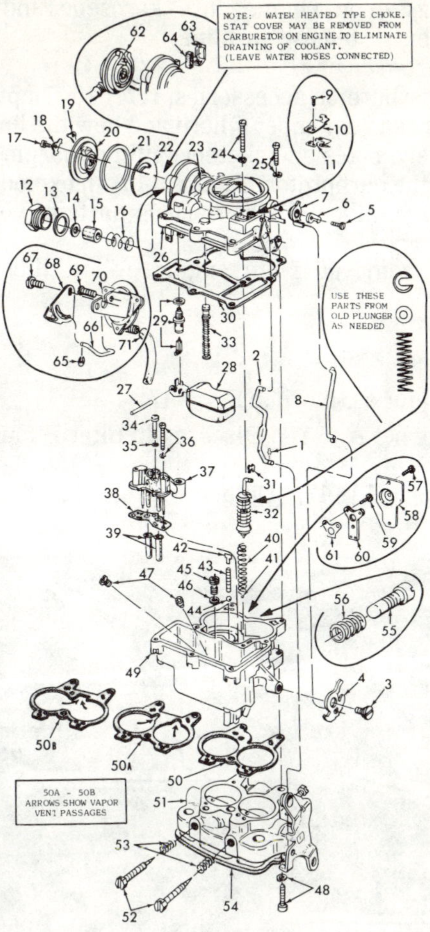

REF.
NO.

1. RETAINER-PUMP ROD LEVER
2. ROD-PUMP
3. SCREW-FAST IDLE CAM
4. CAM-FAST IDLE
5. SCREW-TRIP LEVER
6. LEVER-TRIP
7. LEVER AND COLLAR-CHOKE
8. ROD-CHOKE
9. SCREW-IDLE VENT VALVE
10. SHIELD-IDLE VENT VALVE
11. VALVE-IDLE VENT
12. FITTING-FUEL INLET
13. GASKET-FUEL INLET FITTING
14. GASKET-FUEL FILTER
15. FILTER-FUEL INLET
16. SPRING-FUEL FILTER
17. SCREW-STAT COVER RETAINER
18. RETAINER-STAT COVER-TOOTHED
19. RETAINER-STAT COVER-PLAIN
20. STAT COVER & SPRING ASSY.
21. GASKET-STAT COVER
22. PLATE-CHOKE BAFFLE
23. GASKET-CHOKE HOUSING (NOT SHOWN)
24. SCREW & LOCKWASHER-BOWL COVER (EXTRA LONG)
25. SCREW & LOCKWASHER-BOWL COVER
26. BOWL COVER ASSY.
27. PIN-FLOAT HINGE
28. FLOAT & LEVER ASSY.
29. NEEDLE, SEAT & GASKET ASSY.
30. GASKET-BOWL COVER
31. RETAINER-PUMP PLUNGER
32. PUMP PLUNGER
33. POWER PISTON ASSY.
34. SCREW-VENTURI CLUSTER CENTER
35. WASHER-FIBER-VENTURI CLUSTER CENTER SCREW.
36. SCREW & LOCKWASHER-VENTURI CLUSTER.
37. VENTURI CLUSTER ASSY.
38. GASKET-VENTURI CLUSTER ASSY.
39. INSERT-MAIN WELL
40. SPRING-PUMP RETURN
41. BALL-PUMP INTAKE (USED IN TYPES WITH SEAT IN CASTING).
42. GUIDE-PUMP DISCHARGE BALL
43. SPRING-PUMP DISCHARGE BALL
44. BALL-PUMP DISCHARGE
45. VALVE ASSY.-POWER
46. GASKET-POWER VALVE
47. JETS-MAIN METERING
48. SCREW & LOCKWASHER-THROTTLE BODY TO BOWL
49. BOWL ASSY.-FLOAT
50. GASKET-BODY FLANGE-USED WITH OR WITHOUT IDLE COMPENSATOR (60)

50A GASKET-BODY FLANGE-WITH VAPOR VENT SLOTS-FOR TYPES WITHOUT IDLE COMPENSATOR

50B GASKET-BODY FLANGE-WITH VAPOR VENT OPENING AND IDLE COMPENSATOR PASSAGE-MAY SUB FOR 50 & 50A

51. THROTTLE BODY ASSY.
52. NEEDLE-IDLE ADJUSTING
53. SPRING-IDLE ADJUSTING NEEDLE
54. GASKET-FLANGE
55. NEEDLE-IDLE AIR ADJUSTING (FOR BY-PASS IDLE SYSTEM)
56. SPRING-IDLE AIR ADJUSTING NEEDLE
57. SCREW-COVER IDLE COMPENSATOR
58. COVER-IDLE COMPENSATOR
59. SCREW-IDLE COMPENSATOR VALVE
60. VALVE-IDLE COMPENSATOR
61. GASKET-IDLE COMPENSATOR VALVE
62. STAT COVER & SPRING ASSY. HOT WATER TYPE CHOKE.
63. RETAINER-FILTER
64. FILTER-AIR INTAKE
65. RETAINER-VACUUM CONTROL ROD (2GV)
66. ROD-VACUUM CONTROL (2GV)
67. SCREW-STAT ROD LEVER (2GV)
68. LEVER-STAT ROD (2GV)
69. SCREW-VACUUM CONTROL ATTACHING
70. VACUUM BREAK CONTROL ASSY. (2GV)
71. HOSE-VACUUM BREAK CONTROL (2GV)

Fig. 6-16. Exploded view of Rochester two-barrel carburetor, Models 2G, 2GC, and 2GV.

make up the bath to be used for this task. Tools are listed in Table 6-1.

Preliminary Instructions

Remove air cleaner.

Remove all dirt and grime from exterior of carburetor while carburetor is still mounted on engine; use kerosene and a paintbrush if necessary, but **do not use gasoline.**

Remove carburetor from engine.

For information on carburetor accessories, refer to Chapter 2. If you are not sure you can remember where and how the lines and linkages connect, make notes and sketches in the margin of this book as you remove the carburetor. Figure 6-16, an exploded view, is included to help you identify the various parts of your carburetor.

When carburetor is removed from engine, follow tuneup steps listed below.

To Disassemble

Step 1. Place carburetor as shown in Fig. 6-17.

Step 2. Remove large nut A (6-17). Discard old filter B found beneath nut A.

Step 3. Remove keeper C (6-17). Detach linkage CC at this point.

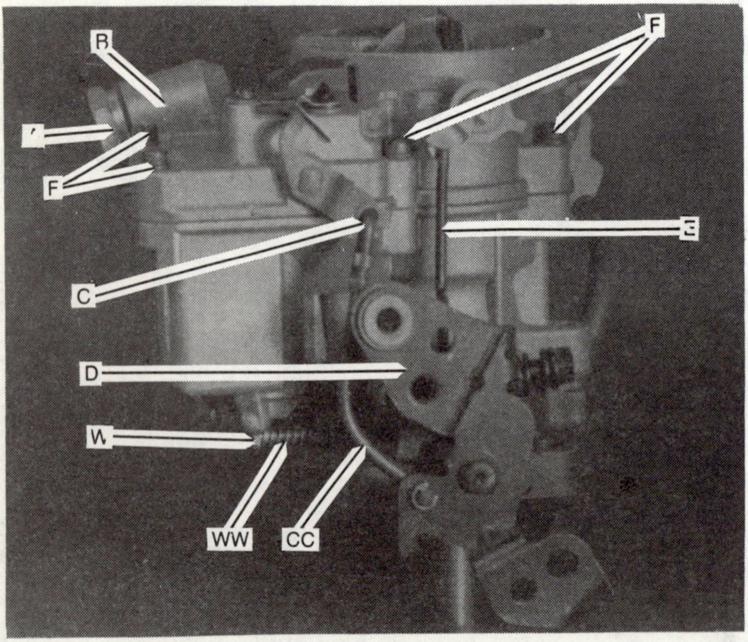

Fig. 6-17. Linkages, filter, cover, and idle-mixture screws.

Step 4. Remove large screw D (6-17). Please pay close attention to the stepdown at the upper end of this linkage so that it can be properly replaced when reassembling. Remove linkage E.

Step 5. Remove the eight screws F (6-17) holding bowl cover to bowl. Lift cover off bowl. Set bowl aside and continue to disassemble cover.

Step 6. Remove pin H (6-18), lift float I off cover.

Step 7. Remove retainer J (6-18). Remove pump plunger assembly K.

Step 8. Remove slotted washer L and spring M from pump plunger. Discard plunger.

Step 9. Remove needle valve N and valve seat O (6-18)—use broad-blade screwdriver. Discard needle and seat.

Step 10. Remove and discard bowl cover gasket. Place bowl cover into carburetor cleaner bath.

Step 11. Remove three jets P (6-19). Place jets in bath.

Step 12. Remove three screws Q (6-19) holding discharge nozzle assembly R. Lift assembly out of bowl. Remove and discard gasket. Place assembly into bath.

Step 13. Remove spring S from pump well T (6-19).

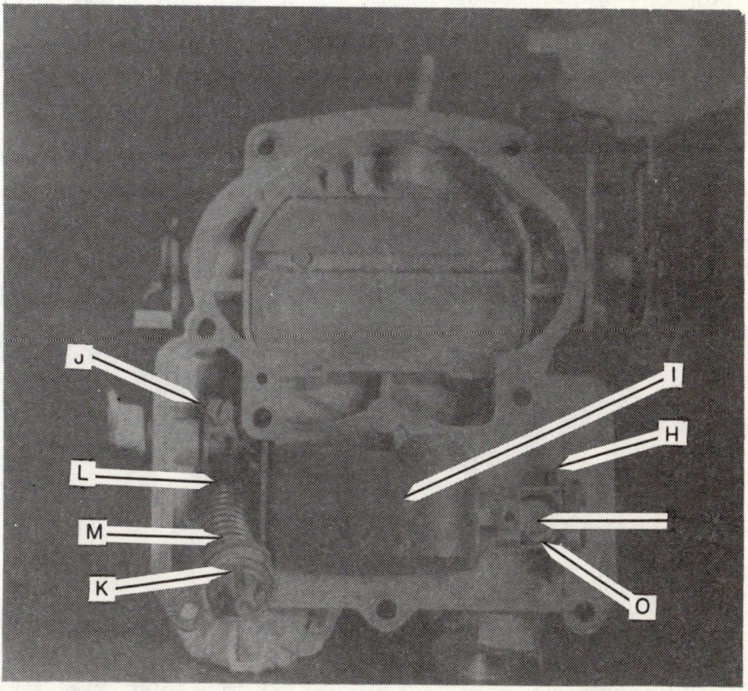

Fig. 6-18. Float, needle and seat, and accelerator pump.

Fig. 6-19. Jets, discharge-nozzle assembly, and accelerator-pump spring.

Step 14. Remove retainer U spring U1, and check ball V from well V1 (6-20) under discharge nozzle. Note size of this check ball, then discard it.

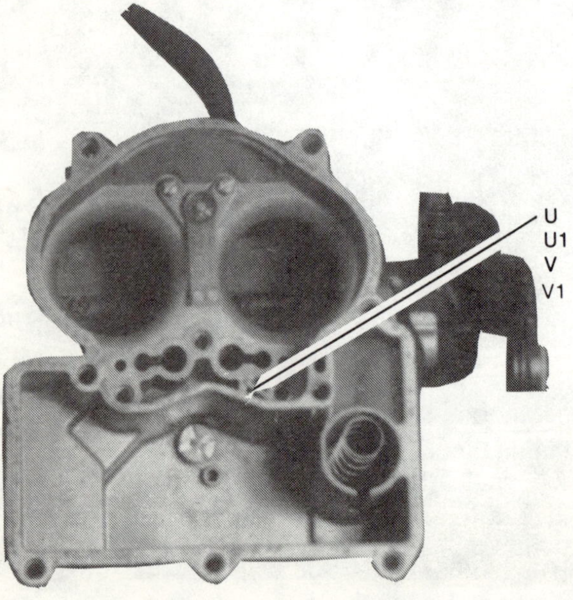

Fig. 6-20. Accelerator-pump check valve.

Step 15. Turn bowl upside down. Remove the two idle mixture screws W and springs WW (6-17).

Step 16. Remove the three screws holding base to bowl. Lift base off bowl and place base in bath.

Step 17. Discard gasket found between base and bowl. Place bowl in bath.

Note: You have completed the disassembly. Allow all parts placed in bath to be properly cleaned in accordance with directions on carburetor-cleaner container before beginning reassembly. The remove all parts from cleaner, rinse with water, and allow to dry naturally. Do not use compressed air.

To Reassemble

Step 18. Place new gasket on bowl and reassemble bowl to base, using three screws.

Step 19. Replace idle-mixture screws W and springs WW (6-12). Tighten screws until they seat lightly, then back them out a complete turn.

Step 20. Turn carburetor assembly upright as shown in (6-21). Use new gaskets. Replace the three jets P (6-21). Tighten jets firmly with broad-blade screwdriver.

Step 21. Replace spring S into accelerator pump well T (6-21).

Step 22. Replace check ball V, spring U1, and retainer U in slotted passage under discharge nozzle assembly.

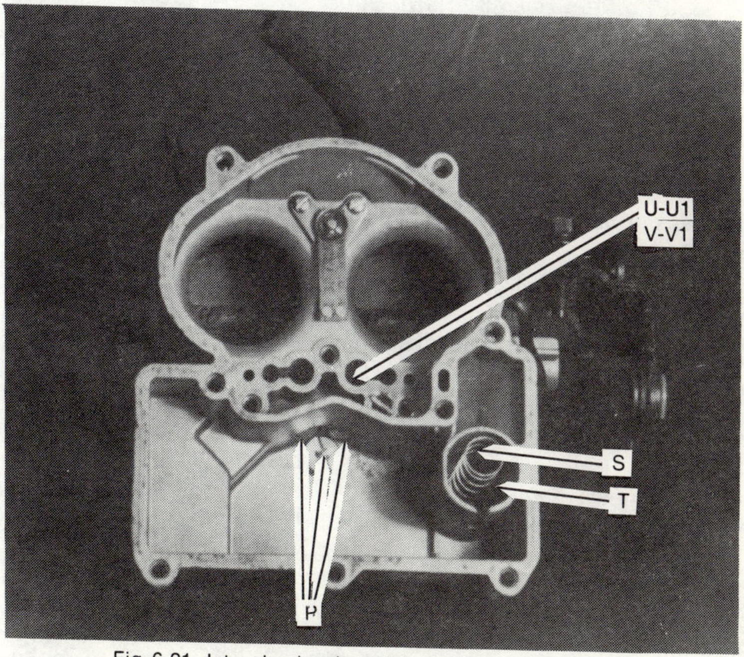

Fig. 6-21. Jets, check-valve retainer, and pump spring.

Fig. 6-22. Nozzle cluster.

Step 23. Install new gasket on discharge nozzle assembly R (6-22). Set assembly in position. Insert three slender screws Q into position and tighten firmly.

Note: The center screw is somewhat different from the other two. It does not use a lockwasher, but does use a fiber gasket. Use a new gasket.

Step 24. Set bowl and base aside and continue assembly by placing a new gasket on bowl cover.

Step 25. Install new needle-valve seat O (6-23). Tighten seat firmly with broad-blade screwdriver.

Step 26. Install spring M and slotted washer L on new accelerator plunger K (6-23).

Step 27. Place plunger assembly in cover (6-23). Secure with new retainer J.

Step 28. Replace needle N into seat O (6-23).

Step 29. Inspect float I for dents or damage. Be sure it does not contain liquid. If float is in good condition, install with pin H (6-23).

Step 30. Adjust float level in accordance with instructions provided with parts kit. Use your best skill to make

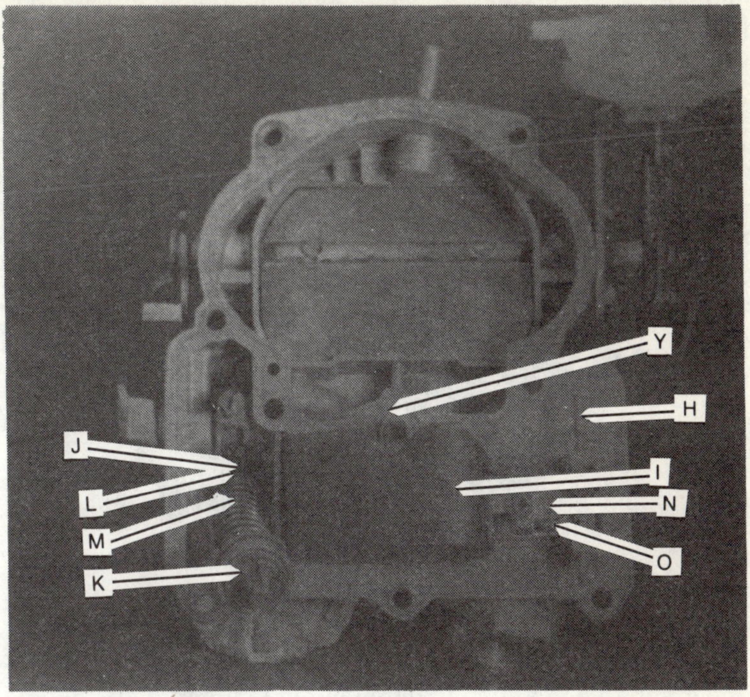

Fig. 6-23. Accelerator pump, float assembly, and power valve.

certain the float level is correct. *This adjustment is very important.*

Step 31. Depress power valve Y (6-25) to be sure it operates freely.

Step 32. Lower cover carefully on bowl. Replace bowl cover screws F (6-24). Run all screws down lightly first, then tighten all screws twice more to be sure they are firm.

Step 33. Replace fast-idle cam linkage E and large-head screw D (6-24).

Step 34. Replace linkage CC and retainer C (6-24).

Step 35. Place new filter B in the opening beneath nut A (6-24). Install nut A.

Step 36. Replace carburetor on engine. Connect all lines and linkages.

Step 37. Start engine; allow it to warm up until choke opens.

Step 38. Adjust idle-speed screw Z (6-24) to desired idle speed.

Step 39. Adjust idle mixture with screws W (6-24). Turn each screw clockwise until engine runs rough; back each screw out counterclockwise until engine runs smooth.

Replace air cleaner. Be sure element is clean.

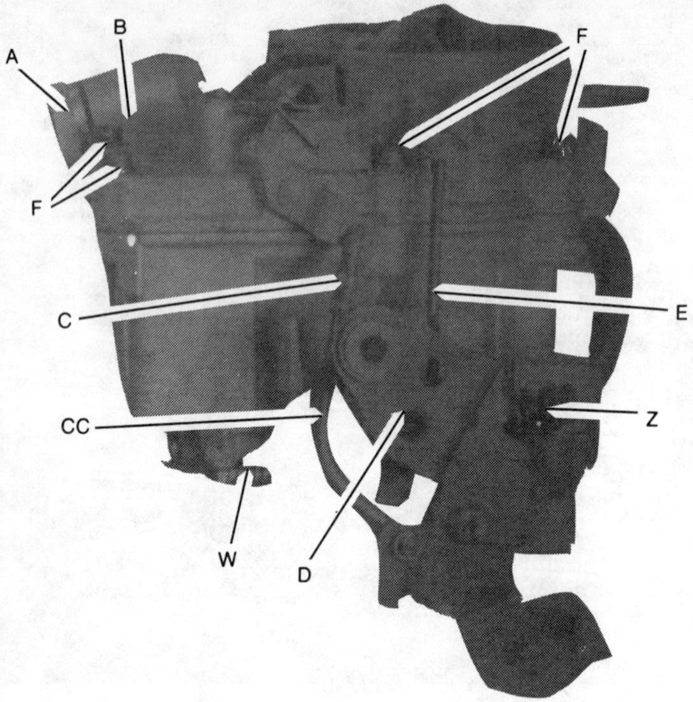

Fig. 6-24. External fittings.

ROCHESTER QUADRAJET (MODEL 4MV)

Please refer to Chapter 1 for a brief description and discussion of carburetor cleaners and the container needed to make up the bath to be used for this task. Tools are listed in Table 6-1.

Preliminary Instructions

Note: The Quadrajet is the most complex of American-built carburetors in popular use, and therefore the most demanding of patience, care, and determination on the part of anyone undertaking carburetor tuneup; however, by following the instructions given here and making all adjustments carefully (there should be a sheet of adjustment instructions in your carburetor parts kit), you can do the job as well as anyone.

Remove air cleaner.

Remove all dirt and grime from exterior of carburetor while carburetor is still mounted on engine; use kerosene and a paintbrush if necessary, but **do not use gasoline.**

Remove carburetor from engine.

For information on carburetor accessories, refer to Chapter 2. If you are not sure you can remember where and how the lines and linkages connect, make notes and sketches in the margin of

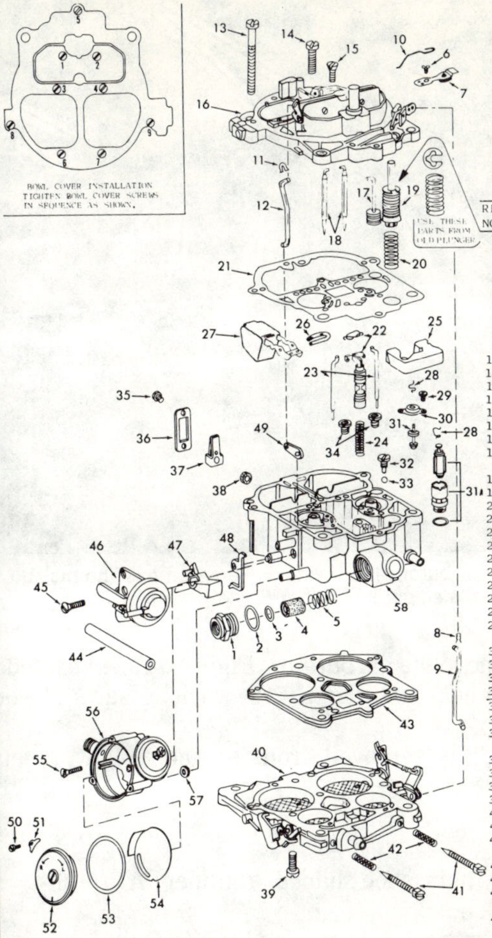

REF.
NO.

1. FILTER NUT - FUEL INLET
2. GASKET - FILTER NUT
3. GASKET - FUEL FILTER
4. FILTER - FUEL INLET
5. SPRING - FUEL FILTER
6. SCREW - IDLE VENT VALVE
7. VALVE - IDLE VENT
8. RETAINER - PUMP ROD
9. ROD - PUMP
10. LEVER - IDLE VENT VALVE
11. RETAINER - CHOKE ROD
12. ROD - CHOKE
13. SCREW - AIR HORN (4)
14. SCREW - AIR HORN (3)
15. SCREW - AIR HORN (2)
16. BOWL COVER ASSY.
17. DASHPOT PISTON & ROD ASSY. 1965-66
18. METERING RODS - SEC. (2)
19. PUMP ASSY.
20. SPRING - PUMP RETURN
21. GASKET - AIR HORN
22. POWER PISTON ASSY. - PRI.
23. METERING RODS - PRI. (2)
24. SPRING - POWER PISTON
25. INSERT - FLOAT BOWL
26. HINGE PIN - FLOAT
27. FLOAT & LEVER ASSY.
28. PULL CLIP - FLOAT NEEDLE
29. SCREW - NEEDLE DIAPHRAGM RETAINER (2)
30. RETAINER - NEEDLE DIAPH.
31. NEEDLE & DIAPHRAGM ASSY.
31A. NEEDLE, SEAT, GASKET ASSY.
32. PLUG - PUMP DISC. BALL
33. BALL - PUMP DISCHARGE
34. JETS - PRIMARY (2)
35. SCREW - IDLE COMPENSATOR COVER (2)
36. COVER - IDLE COMPENSATOR
37. IDLE COMPENSATOR ASSY.
38. GASKET-IDLE COMPENSATOR
39. SCREW - THROTTLE BODY (3)
40. THROTTLE BODY ASSY.
41. NEEDLES - IDLE ADJ. (2)
42. SPRINGS - IDLE ADJ. NEEDLE (2)
43. GASKET - THROTTLE BODY
44. HOSE - VACUUM 4MV
45. SCREW - VACUUM BREAK CONTROL BRACKET ATTACHING
46. VACUUM BREAK CONTROL & BRACKET ASSY. 4MV
47. CAM - FAST IDLE
48. LEVER - SEC. LOCKOUT 4MV
49. LEVER - INTERMEDIATE CHOKE
50. SCREW - STAT RETAINER (3) 4MC
51. RETAINERS - STAT COVER (3) 4MC
52. STAT COVER & SPRING ASSY. 4MC
53. GASKET - STAT COVER 4MC
54. PLATE - CHOKE BAFFLE 4MC
55. SCREW - STAT HOUSING ATT.
56. HOUSING - CHOKE & VACUUM BREAK ASSY. 4MC
57. GASKET - CHOKE HOUSING 4MC
58. BOWL ASSY. - FLOAT

Fig. 6-25. Exploded view of a Rochester Quadrajet four-barrel carburetor, Model 4MV.

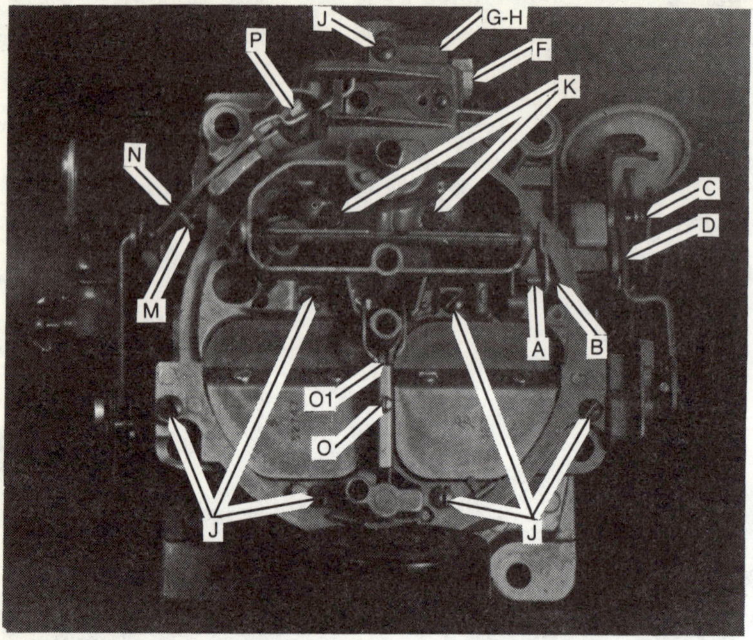

Fig. 6-26. Top view of carburetor, showing linkage-removal points, bowl-cover screws, filter, and related parts.

this book as you remove the carburetor. Figure 6-25, an exploded view, is included to help you identify the various parts of your carburetor.

When carburetor is removed from engine, follow tuneup steps listed below.

To Disassemble

Step 1. Remove horseshoe-shaped retainer A (6-26) and release linkage B.
Step 2. Remove retainer C (6-26) and linkage D.
Step 3. Remove large nut F (6-26) and spring G. Discard filter H from within spring G.
Step 4. If your carburetor has one, remove baffle H (6-26).
Step 5. Remove the remaining seven bowl cover screws J and K (6-26). Note that two screws K are situated within throat of carburetor.
Step 6. Remove horseshoe-shaped retainer M (6-26) and detach linkage N.
Step 7. Remove screw O (6-26); lift out the secondary metering rod assembly O1; place assembly in bath.
Step 8. Lift bowl cover off bowl; discard gasket; place bowl cover in bath.

Fig. 6-27. Accelerator-pump spring, vacuum pistons, metering-rod assembly, and plastic molding.

Step 9. Remove accelerator plunger and spring Q (6-27); place spring in bath.

Step 10. Remove plastic molding R (6-27).

Step 11. Remove vacuum pistons S and metering rod assembly T (6-27). Remove rods by pulling upward firmly. Place assembly in bath.

Step 12. Lift float U and needle V (6-28) from bowl. Discard needle.

Step 13. Remove screw W from alongside accelerator pump well X (6-28). Underneath this screw is a steel check ball Y. Note size of ball, then discard it.

Step 14. Remove needle-valve seat Z (6-28). Discard it.

Step 15. Remove screw AA (6-28); remove choke vacuum assembly BB. Do not disassemble choke vacuum assembly unless you are particularly adept at puzzles.

Step 16. Turn carburetor assembly upside down. Use Phillips screwdriver to remove base-retaining screws CC (6-29). Separate base from bowl. Discard gasket. Place bowl in bath.

Step 17. Remove idle-mixture screws DD and springs EE (6-29). Place screws and springs in bath.

Step 18. Place base in bath.

Fig. 6-28. Accelerator-pump check valve, needle and seat, and float.

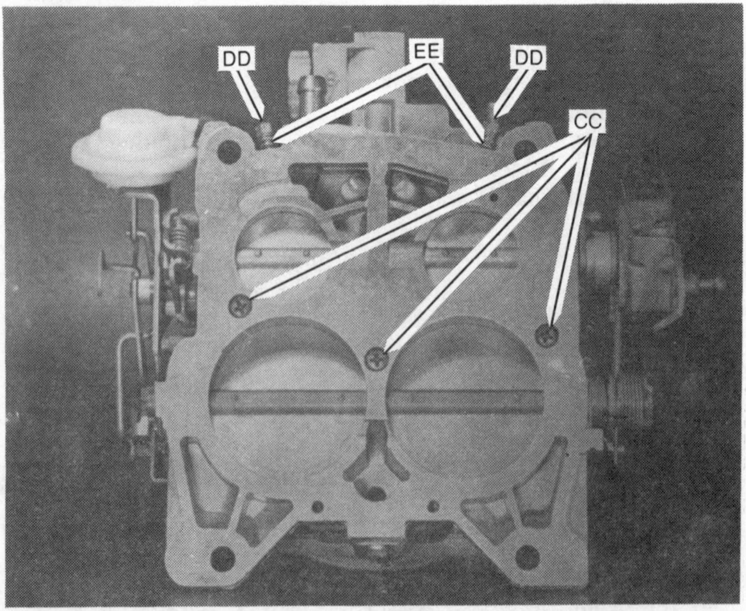

Fig. 6-29. Base-to-bowl and idle-mixture screws.

Note: You have completed the diassembly. Allow all parts placed in bath to be properly cleaned in accordance with directions on carburetor-cleaner container before beginning reassembly. Then remove all parts from cleaner, rinse with water, and allow to dry naturally. Do not use compressed air.

To Reassemble

Step 19. Replace idle-mixture screw springs EE and screws DD (6-29). Tighten screws lightly until needles seat in base, then back each screw out a complete turn.

Note: Inspect Welch plugs. If evidence of fuel leakage is found, replace with new plugs and *O*-rings (6-30).

Step 20. Place new gasket on base and reassemble bowl to base with Phillips screws CC (6-29).

Step 21. Assemble choke-vacuum assembly BB to bowl (6-31) with screw AA.

Step 22. Install new check ball Y in hole Y1 (6-32). Replace retainer screw W with pin end on top of check ball. Tighten firmly.

Step 23. Install new needle-valve seat Z (6-32). Tighten firmly with broad-blade screwdriver.

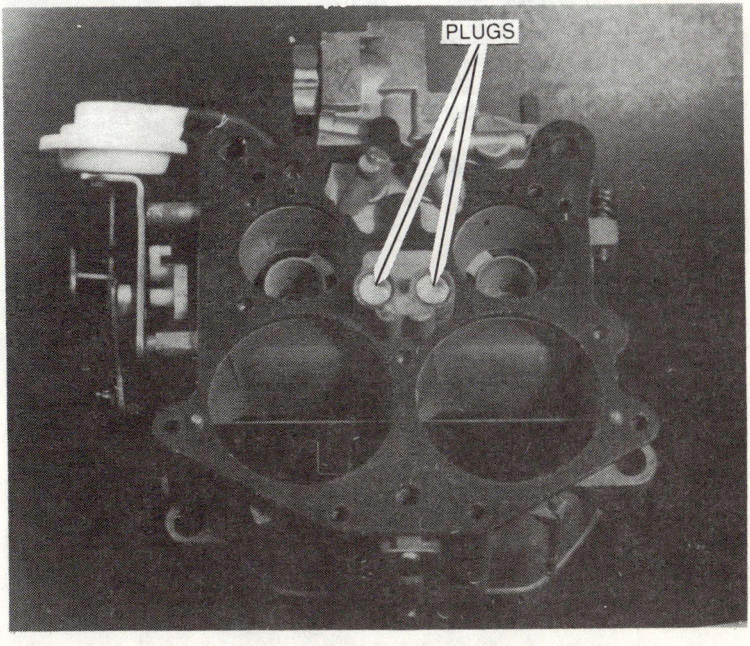

Fig. 6-30. Welch plugs. If leaking replace with plug kit or seal with epoxy.

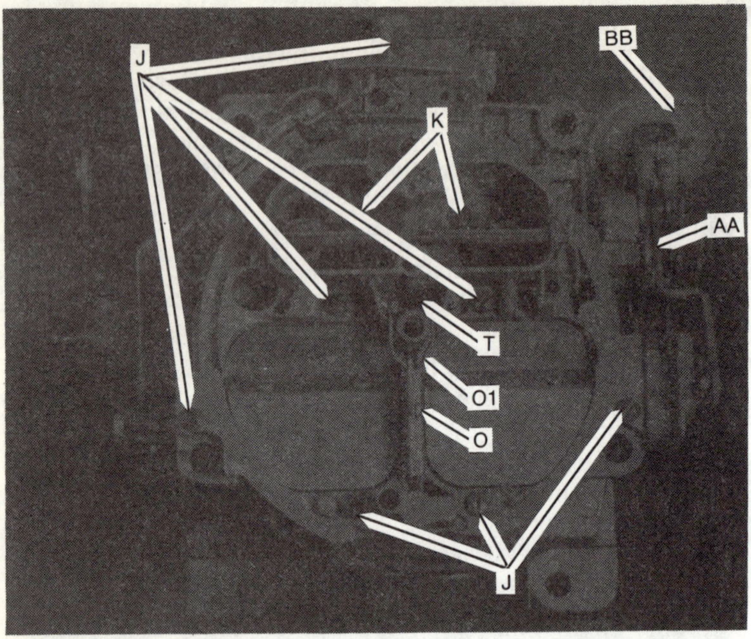

Fig. 6-31. Choke-vacuum assembly, linkage, and metering assembly.

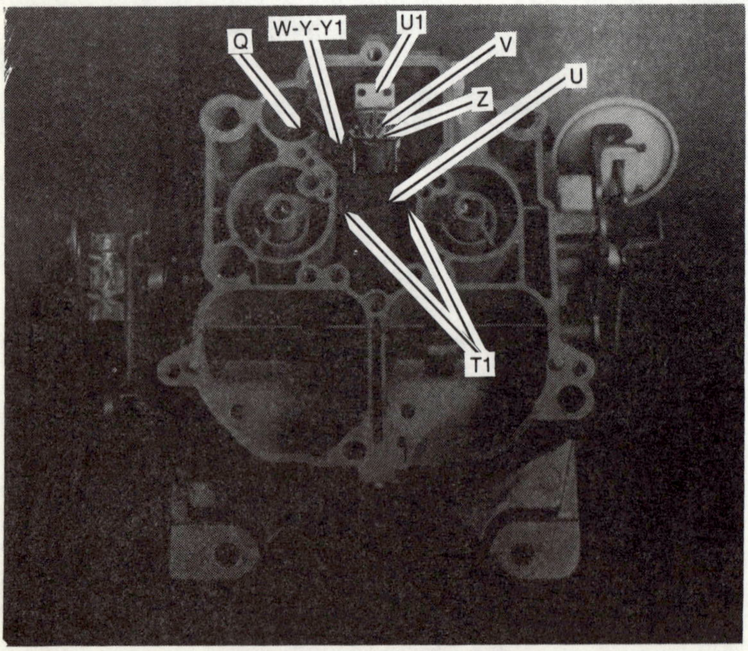

Fig. 6-32. Float and related parts.

Step 24. Tighten metering jets T1 (6-32) in bottom of carburetor bowl. If it appears necessary, remove these jets and install with new gaskets furnished in parts kit.

Step 25. Place needle V (6-32) in position on end of float U. Lower float and needle into position as a unit.

Step 26. With float U in position, use a pencil or similar instrument to press down lightly on assembly U1 (6-32) and release to test freedom of float action.

Step 27. Make float-level adjustment in accordance with instructions in parts kit. *Be accurate.*

Step 28. Replace vacuum pistons S (6-33) and metering rod assembly. This task can be quite tricky. The recommended procedure is to release the tiny tension spring at top end of the assembly while aligning the metering rods into jets at bottom of bowl. Once the assembly is in position, push it down firmly, then release it. Assembly should now move up and down freely. Attach tension springs to tops of metering rods.

Step 29. Place plastic molding R (6-33) in position.

Step 30. Place new bowl cover gasket in position.

Step 31. Remove slotted washer P1 (6-34) and spring P2 from accelerator pump plunger and install them on new plunger from parts kit.

Fig. 6-33. Vacuum pistons, metering rods, plastic molding, and accelerator-pump spring.

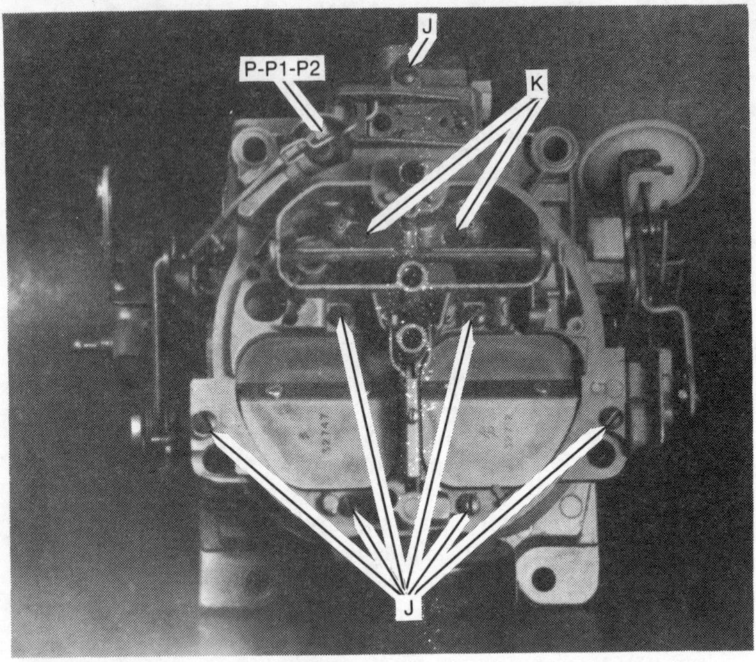

Fig. 6-34. Bowl-cover screws and external pump parts.

Step 32. Install return spring Q (6-33) into pump well.

Step 33. Insert accelerator pump plunger P (6-34) into well.

Step 34. Hold accelerator pump in position with one finger; turn bowl cover over and carefully lower it into position on bowl. Align metering rods T (6-33) with proper holes in new bowl cover gasket.

Step 35. Replace bowl cover screws J and K (6-34); tighten firmly.

Step 36. Install secondary metering rod assembly O1 with screw O (6-35). Be sure rods line up with jets in bottom of bowl—if properly aligned, there should be no binding of the rods.

Step 37. Replace linkage N (6-35) and retainer M.

Step 38. Replace linkage D (6-35) and retainer C.

Step 39. Replace linkage B (6-35) and retainer A.

Note: Lower end of linkage must be hooked into lever at bottom before top end is attached.

Step 40. Install new filter H (6-35) into spring G and replace both under large nut F (over new gasket). Tighten firmly.

Step 41. Replace carburetor on engine. Attach all linkages and fittings.

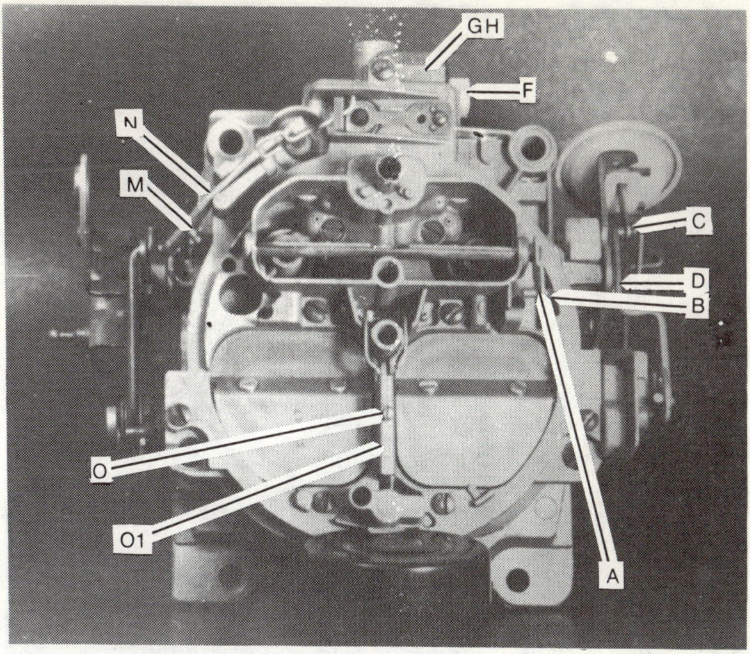

Fig. 6-35. Secondary metering-rod assembly, filter, and linkages.

Step 42. Start engine. Allow it to warm up until choke valve is fully open.

Step 43. Adjust idle-speed screw EE (6-36) until preferred idle speed is obtained.

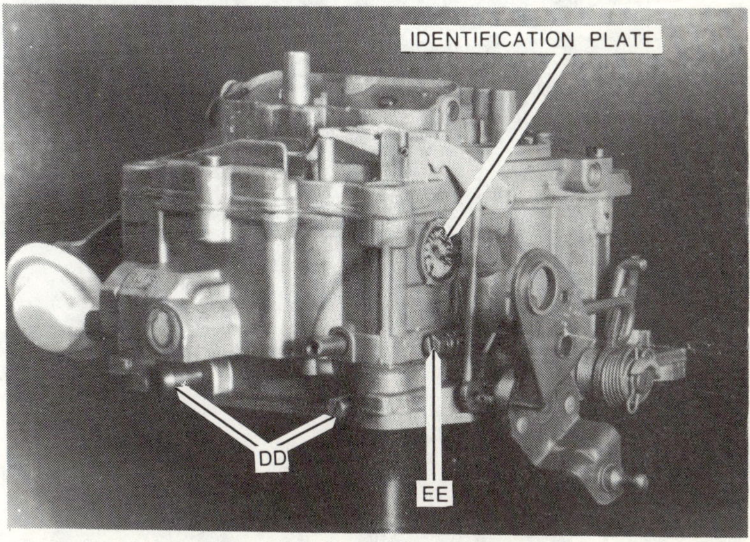

Fig. 6-36. Idle-speed and mixture screws.

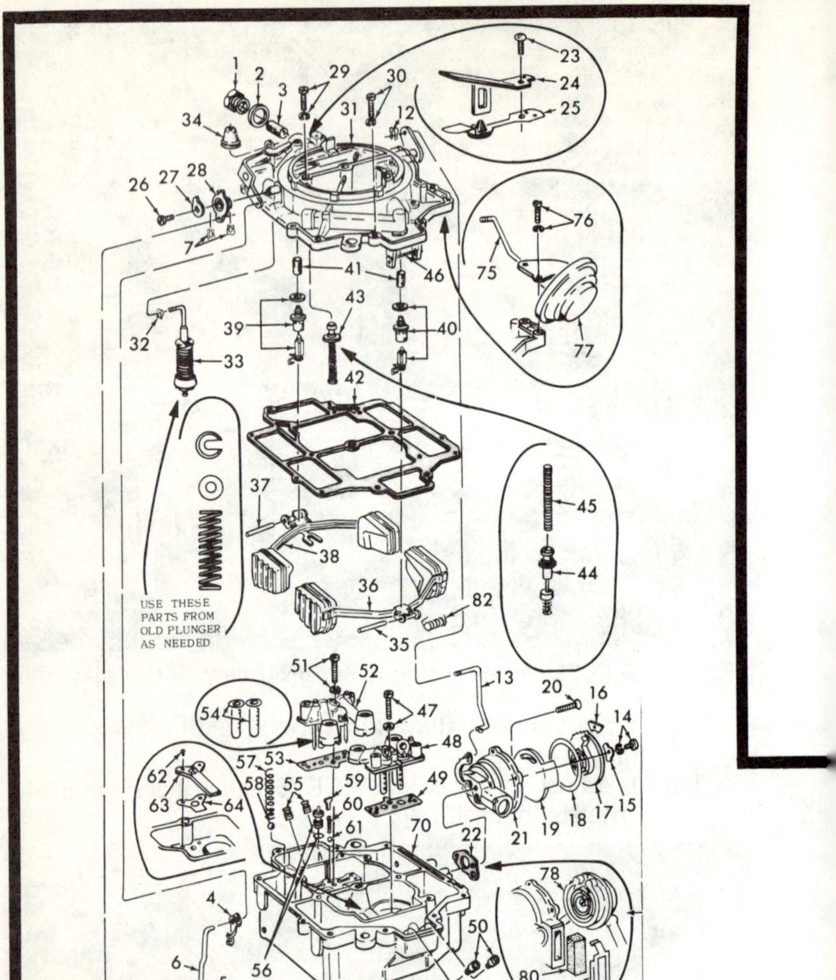

Fig. 6-37. Exploded view of a Rochester four-barrel carburetor, Model 4GC.

Step 44. Adjust idle mixture with screws DD (6-36). Turn screws clockwise until engine begins to run rough; then back out counterclockwise just enough to obtain smooth idle. Alternate adjustments between the two screws until you are satisfied with engine idle.

Replace air cleaner. Be sure element is clean.

ROCHESTER FOUR-BARREL (MODEL 4GC)

Please refer to Chapter 1 for a brief description and discussion of carburetor cleaners and the container needed to make up the bath to be used for this task. Tools are listed in Table 6-1.

Preliminary Instructions

Remove air cleaner.

Remove all dirt and grime from exterior of carburetor while carburetor is still mounted on engine; use kerosene and a paintbrush if necessary, but **do not use gasoline.**

Remove carburetor from engine.

For information on carburetor accessories, refer to Chapter 2. If you are not sure you can remember where and how the lines and linkages connect, make notes and sketches in the margin of this book as you remove the carburetor. Figure 6-37, an exploded view, is included to help you identify the various parts of your carburetor.

REF. NO.		REF. NO.	
1.	FITTING-FUEL INLET	45.	SPRING-POWER PISTON
2.	GASKET-FUEL INLET FITTING	46.	SPRING-FLOAT BALANCE
3.	SCREEN-FUEL INLET	47.	SCREW & LOCKWASHER-SECONDARY VENTURI CLUSTER
4.	RETAINER-PUMP ROD UPPER	48.	SECONDARY VENTURI CLUSTER
5.	RETAINER-PUMP ROD LOWER	49.	GASKET-SEC. VENTURI CLUSTER
6.	ROD-PUMP	50.	JETS-SECONDARY MAIN METERING
7.	RETAINER-CHOKE ROD UPPER	51.	SCREW & LOCKWASHER-PRIMARY VENTURI CLUSTER
8.	RETAINER-CHOKE ROD LOWER	52.	PRIMARY VENTURI CLUSTER
9.	ROD-CHOKE	53.	GASKET-PRI. VENTURI CLUSTER
10.	SCREW-FAST IDLE CAM	54.	INSERTS-MAIN WELL
11.	CAM-FAST IDLE	55.	JETS-PRIMARY MAIN METERING
12.	RETAINER-INTERMEDIATE CHOKE ROD	56.	VALVE & GASKET ASSY.-POWER
13.	ROD-INTERMEDIATE CHOKE	57.	SPRING-PUMP RETURN
14.	SCREW AND LOCKWASHER-STAT RETAINER	58.	BALL-PUMP INTAKE (USED IN TYPES WITH SEAT IN CASTING
15.	RETAINER-STAT COVER PLAIN	59.	GUIDE-PUMP DISCHARGE BALL
16.	RETAINER-STAT COVER TOOTHED	60.	SPRING-PUMP DISCHARGE BALL
17.	STAT COVER AND SPRING ASSY.	61.	BALL-PUMP DISCHARGE
18.	GASKET-STAT COVER	62.	SCREW-IDLE COMPENSATOR VALVE
19.	PLATE-CHOKE BAFFLE	63.	VALVE-IDLE COMPENSATOR
20.	SCREW-CHOKE HOUSING	64.	GASKET-IDLE COMPENSATOR VALVE
21.	HOUSING-CHOKE & PISTON ASSY.	65.	SCREW & LOCKWASHER-THROTTLE BODY ATTACHING (CENTER HOLE)
22.	GASKET-CHOKE HOUSING	66.	SCREW & LOCKWASHER-THROTTLE BODY ATTACHING
23.	SCREW-IDLE VENT VALVE	67.	THROTTLE BODY ASSEMBLY
24.	SHIELD-IDLE VENT VALVE	68.	GASKET-BODY FLANGE
25.	VALVE-IDLE VENT	68A	GASKET-BODY FLANGE WITH VAPOR VENT SLOTS. WILL REPLACE #68
26.	SCREW-TRIP LEVER	69.	AUXILARY THROTTLE VALVE ASSY.
27.	LEVER-TRIP	70.	BOWL ASSEMBLY-FLOAT
28.	LEVER AND COLLAR-CHOKE	71.	NEEDLES-IDLE ADJUSTING
29.	SCREW & LOCKWASHER-BOWL COVER	72.	SPRINGS-IDLE ADJ. NEEDLE
30.	SCREW & LOCKWASHER-BOWL COVER	73.	NEEDLE-IDLE AIR ADJUSTING (FOR BY-PASS IDLE SYSTEM)
31.	BOWL COVER ASSEMBLY	74.	SPRING-IDLE AIR ADJ. NEEDLE
32.	RETAINER-PUMP PLUNGER	75.	ROD-VACUUM CONTROL
33.	PUMP PLUNGER	76.	SCREW & LOCKWASHER-VACUUM CONTROL ATTACHING
34.	BOOT-PUMP PLUNGER	77.	VACUUM BREAK CONTROL ASSEMBLY
35.	PIN-SECONDARY FLOAT HINGE	78.	STAT COVER & SPRING ASSEMBLY HOT WATER TYPE CHOKE
36.	FLOAT & LEVER ASSY. SECONDARY	79.	RETAINER-FILTER
37.	PIN-PRIMARY FLOAT HINGE	80.	FILTER-AIR INTAKE
38.	FLOAT & LEVER ASSY. PRIMARY	81.	THROTTLE RETURN CHECK ASSEMBLY
39.	NEEDLE, SEAT & GASKET ASSY. PRIMARY	82.	TORSION SPRING-FLOAT
40.	NEEDLE, SEAT & GASKET ASSY. SECONDARY		
41.	SCREEN-NEEDLE & SEAT STRAINER		
42.	GASKET-BOWL COVER		
43.	POWER PISTON ASSEMBLY		
44.	POWER PISTON WITH FLOAT ASSIST SPRING ASSEMBLY		

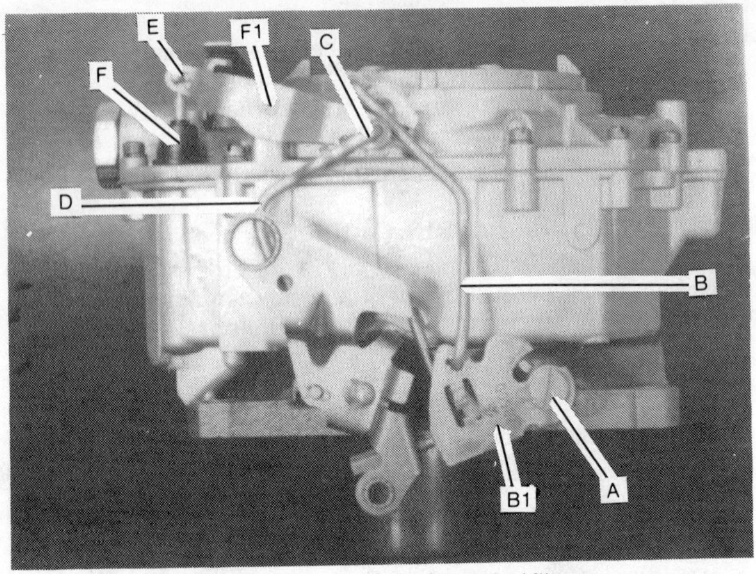

Fig. 6-38. External linkages and fuel filter.

When carburetor is removed from engine, follow tuneup steps listed below.

To Disassemble

Step 1. Remove screw A (6-38), linkage B, and lever B1. Place lever and linkage in bath.

Step 2. Remove spring clip C (6-38) and release linkage D at point C1.

Step 3. Remove clip E (6-38) and release accelerator pump plunger F from lever F1.

Step 4. Remove clip G (6-39) and linkage H. Place linkage in bath.

Step 5. Remove three screws I (6-39), retainers I1, and choke housing J. Discard gasket.

Step 7. From cover, remove accelerator pump assembly L (6-40). Discard boot M (6-40).

Step 8. Remove slotted washer L1 (6-40) and coil spring L2 from accelerator-pump plunger. Place washer and spring in bath. Discard plunger.

Step 9. Remove large nut N (6-40). Discard filter N1 and gasket from beneath nut. Place nut in bath.

Step 10. Remove two float-hinge pins O (6-40) and lift out both floats P. Do *not* place floats in bath.

Step 11. Remove two needle valves Q1 (6-40) from seats Q; remove seats (use broad-blade screwdriver). Discard seats, gaskets, and valves.

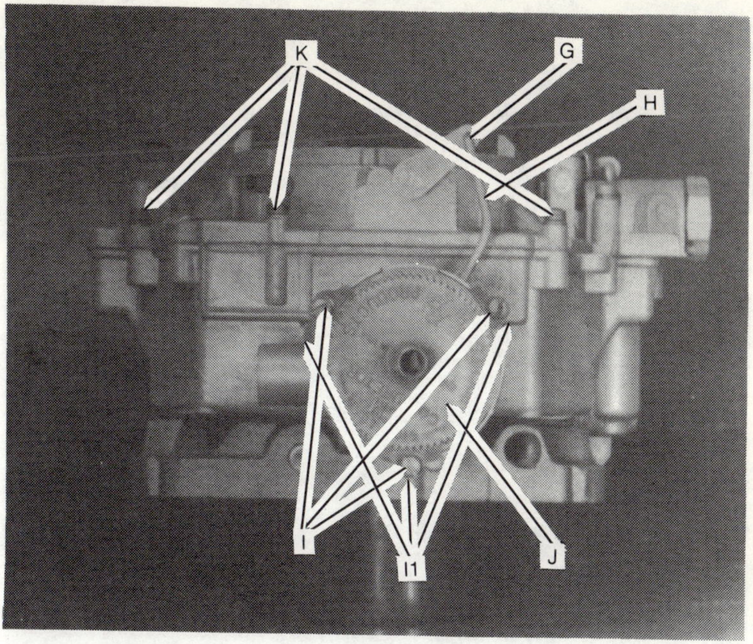

Fig. 6-39. Cover screws, linkage, choke assembly, and filter.

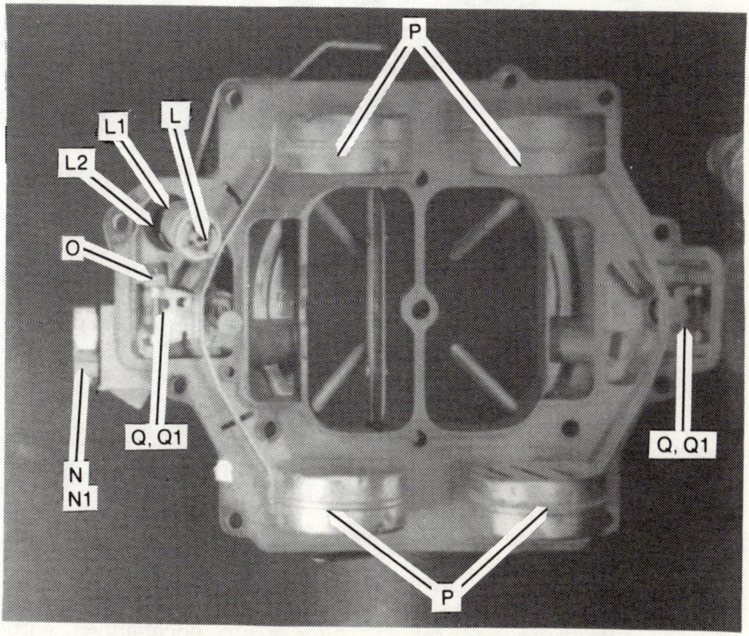

Fig. 6-40. Floats, needles and seats, and accelerator pump.

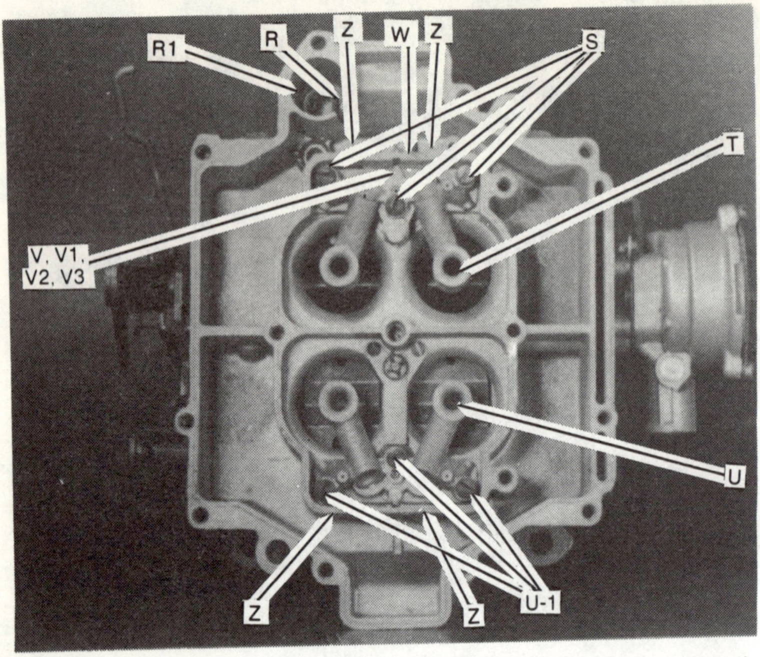

Fig. 6-41. Accelerator-pump spring and well, primary and secondary venturi clusters, and jet.

Step 12. Place bowl cover in bath.

Step 13. Remove coil spring R (6-41) from well. Place spring in bath.

Step 14. Remove three screws S (6-41); lift out primary venturi cluster T and place it in bath.

Step 15. Remove three screws U1 (6-41); lift out secondary venturi cluster U and place it in bath.

Step 16. In center of bowl, beneath primary venturi cluster, is well V (with a metal retainer V1 across it). Use needle-nosed pliers to lift out retainer; remove coil spring V2 and steel ball V3 from well. Discard ball. Place spring in bath.

Step 17. Use broad-blade screwdriver to remove jet W (6-41). Place jet in bath.

Step 18. Remove two idle-mixture adjustment screws X (6-42) and springs X1. Place screws and springs in bath.

Step 19. Remove four screws Y (6-42). Note their sizes.

Step 20. Separate bowl from base. Discard gasket. Place bowl and base in bath.

Note: You have completed the disassembly. Allow all parts placed in bath to be properly cleaned in accordance with directions on carburetor-cleaner container before beginning reassembly. Then remove all parts from bath, rinse with water, and allow to dry naturally. Do not use compressed air.

To Reassemble

Step 21. Use new gasket. Assemble carburetor bowl to base. Use four screws Y (6-42)—remember the larger one fits into hole in center of base. Tighten firmly.

Step 22. Install idle-mixture adjustment screws X (6-42) into springs X1 and into carburetor base. Run screws in until they seat lightly; then back each out 1¼ turns.

Step 23. Tighten four jets Z (6-43) in bottom of bowl.

Step 24. Install jet W (6-43); use broad-blade screwdriver. Tighten firmly.

Step 25. Drop new steel ball V3 (6-43) into well V; place coil spring V2 on top of ball; place *T*-shaped retainer V1 into center of spring and push it down into position—tap it with screwdriver handle if necessary.

Step 26. Use new gasket. Install primary venturi cluster assembly T (6-43) with three screws S. Tighten firmly.

Step 27. Use new gasket. Install secondary venturi cluster assembly U (6-43) with three screws U1. Tighten firmly.

Step 28. Drop coil spring R (6-43) into well R1.

Step 29. Place new rubber boot M (6-43) in position.

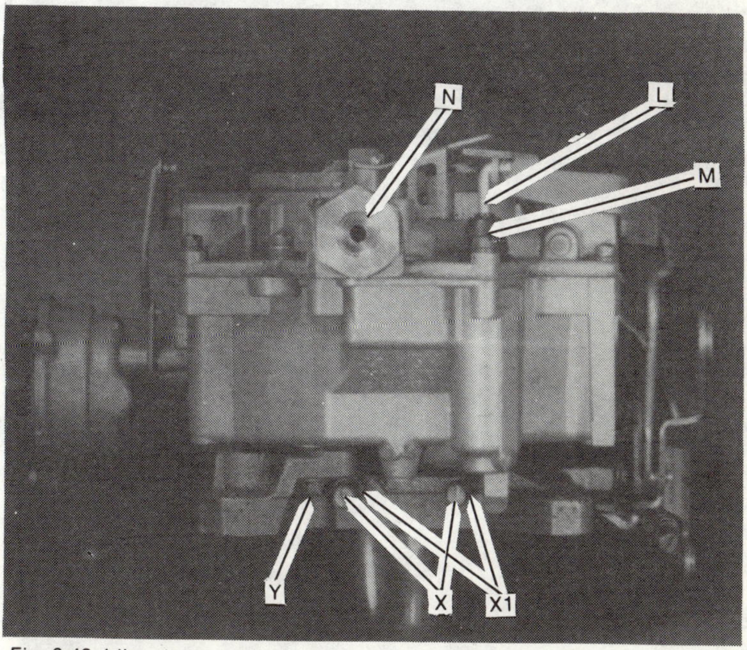

Fig. 6-42. Idle-mixture screws, base-to-bowl screws, accelerator-pump boot, and filter.

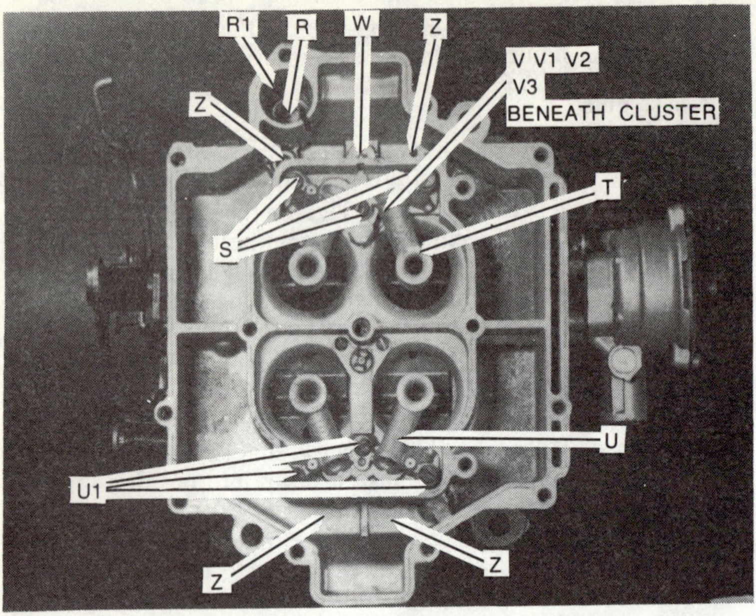

Fig. 6-43. Jets, check ball, primary and secondary venturi clusters, and pump spring.

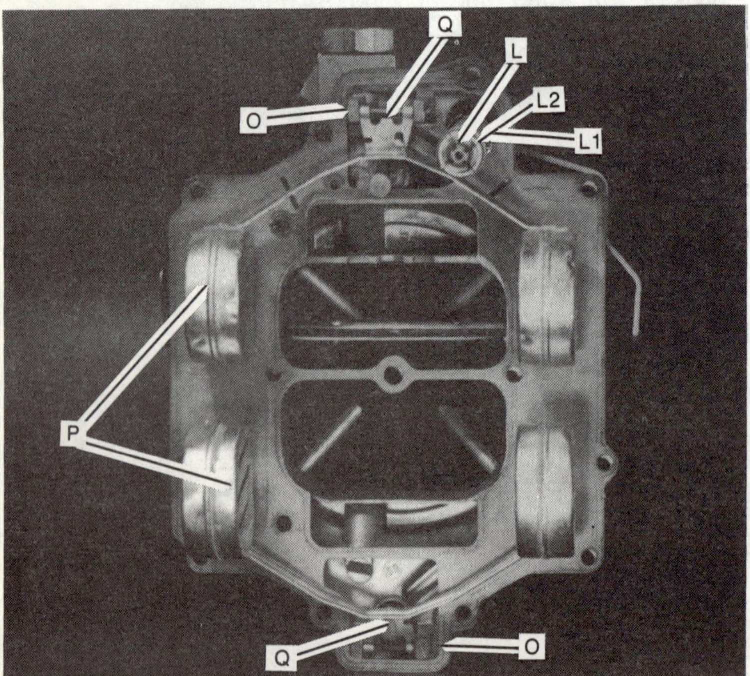

Fig. 6-44. Float assemblies and accelerator-pump plunger.

Step 30. Use new gaskets. Install two new needle-valve seats Q (6-44). Tighten firmly.

Step 31. Drop new needle valves Q (6-44) into seats, point first.

Step 32. Place new bowl cover gasket into position on bowl.

Step 33. Install floats P (6-44) with hinge pins O.

Step 34. Adjust float level in accordance with instruction sheet included in your parts kit. *Be accurate.*

Step 35. Over new accelerator pump plunger L (6-44) place coil spring L2, then slotted washer L1, cup side seated over end of spring.

Step 36. Insert plunger through hole in cover and through boot M.

Step 37. Lower cover into position (carefully) over bowl. Insert and tighten thirteen screws K (6-45).

Step 38. Attach linkage H (6-47) with clip G.

Step 39. Use new gasket. Install choke housing J (6-45) with retainers I1 and screws I (not tightened yet). Turn housing clockwise to engage choke lever; align mark on housing with mark on casting. Tighten screws.

Step 40. Attach accelerator pump plunger F (6-46) to lever F1 with clip E.

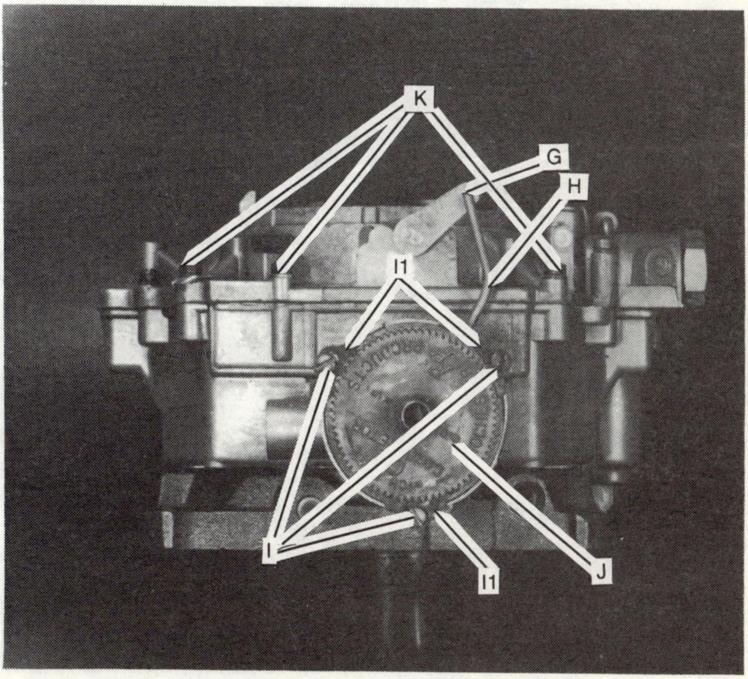

Fig. 6-45. Choke housing, link, and cover screws.

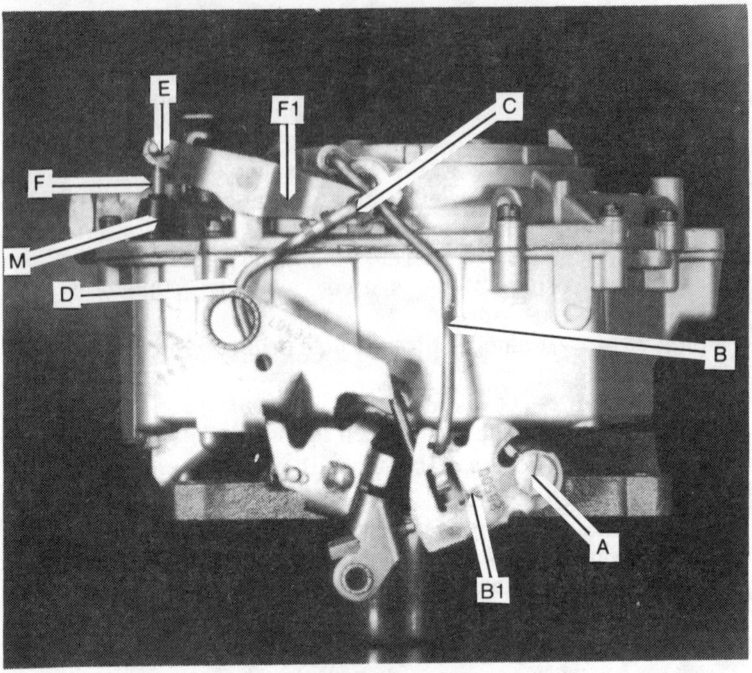

Fig. 6-46. External linkage.

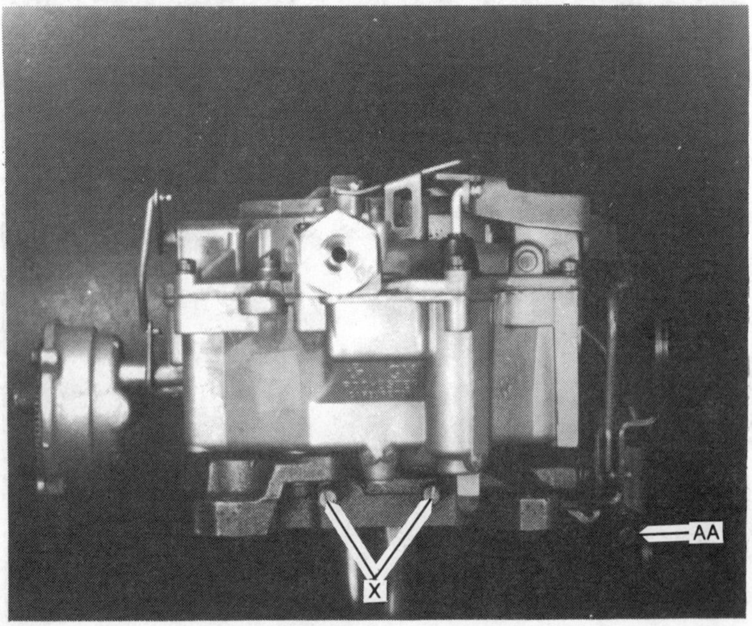

Fig. 6-47. Idle-speed and mixture screws.

Step 41. Attach linkage D (6-46) with clip C.

Step 42. Attach linkage B (6-46) to lever B1. Tighten screw firmly.

Step 43. Install carburetor on engine. Be sure fuel line is tightly fitted and that all linkages are properly attached.

Step 44. Start engine. Allow it to warm up until choke butterfly valve is in fully open position.

Step 45. Adjust idle to desired speed with screw AA (6-47).

Step 46. Adjust idle mixture by turning screws X (6-47)—one at a time—clockwise until it runs smoothly. Adjust screws alternately until you achieve the smoothest idle.

Replace the air cleaner. Be sure element is clean.

Chapter 7

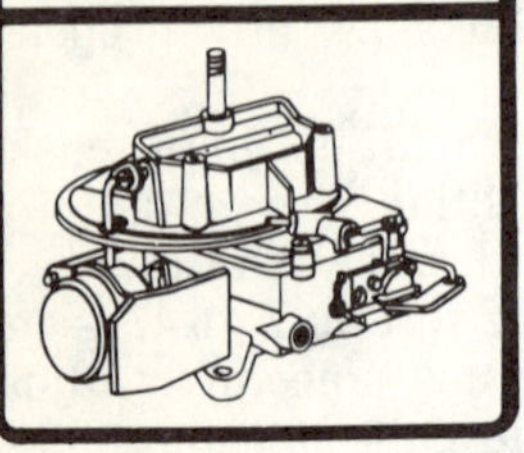

Stromberg

Stromberg is a name that has been around for a long time. Some readers might remember the Stromberg 97s (adjustable main jets and all) fitted to flat-head Fords. Today the firm, part of Bendix, is no longer in the OEM market. The last original equipment carburetors were on 1967 Chrysler products and 1968 GMC trucks. Rumor has it that Stromberg bowed out of this lucrative market rather than retool to meet emission-control requirements. Nevertheless, Stromberg is still very much in business, building replacement carburetors and parts.

STROMBERG TWO-BARREL (MODEL WW)

Please refer to Chapter 1 for a brief description and discussion of carburetor cleaners and the container needed to make up the bath to be used for this task. Tools are listed in Table 7-1.

Table 7-1. Tools Required for Stromberg Carburetor Overhaul.

Screwdriver with a broad blade (approx. 7/16 in.).
Needle-nosed pliers.
Open-end or box-end wrench, 9/16 in.
Adjustable wrench with at least a ¾ in. span.
Open-end, box-end, or socket wrench of the proper size to remove the nuts or bolts that hold your carburetor to the manifold.

Preliminary Instructions

Remove air cleaner.

Remove all dirt and grime from exterior of carburetor while carburetor is still mounted on engine; use kerosene and a paintbrush if necessary, but **do not use gasoline**.

Remove carburetor from engine.

For information on carburetor accessories, refer to Chapter 2. If you are not sure you can remember where and how the lines and linkages connect, make notes and sketches in the margin of this book as you remove the carburetor. Figure 7-1, an exploded view, is included to help you identify the various parts of your carburetor.

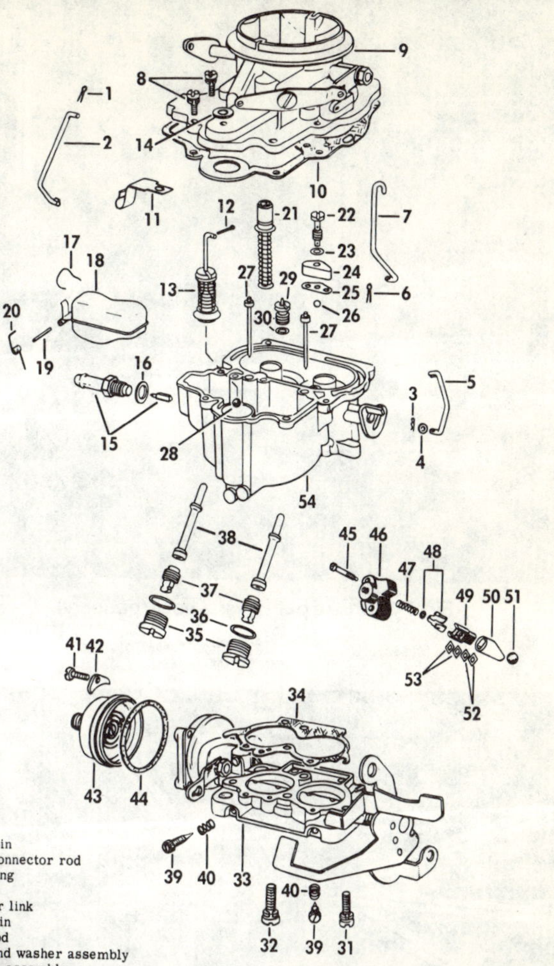

REF. NO.	
1	Cotter pin
2	Choke connector rod
3	Pin spring
4	Washer
5	Unloader link
6	Cotter pin
7	Pump rod
8	Screw and washer assembly
9	Air horn assembly
10	Air horn gasket
11	Cable clip
12	Cotter pin
13	Pump plunger assembly
14	Pump plunger washer
15	Needle and seat assembly
16	Needle seat gasket
17	Float pin spring
18	Float assembly
19	Float pin
20	Float spring
21	Vacuum power piston assembly
22	Pump discharge nozzle screw
23	Pump discharge nozzle screw gasket
24	Pump discharge nozzle
25	Pump discharge nozzle gasket
26	Pump outlet check ball
27	Idle tube
28	Pump inlet check ball
29	Power by-pass jet
30	Power by-pass jet gasket
31	Screw and washer assembly (short)
32	Screw and washer assembly (long)
33	Throttle body assembly
34	Main body gasket
35	Main discharge jet plug
36	Main jet plug gasket
37	Main metering jet
38	Main discharge nozzle
39	Idle needle valve
40	Idle needle valve spring
41	Screw (self-tapping)
42	Thermostat cover retainer
43	Thermostat and cover assembly
44	Thermostat cover gasket
45	Starter switch cover screw
46	Starter switch cover
47	Switch return spring
48	Switch contact spring
49	Switch guide block
50	Switch plunger
51	Switch ball
52	Switch adjustment washer (thick)
53	Switch adjustment washer (thin)
54	Main body

Fig. 7-1. Exploded view of Stromberg two-barrel carburetor, Model WW.

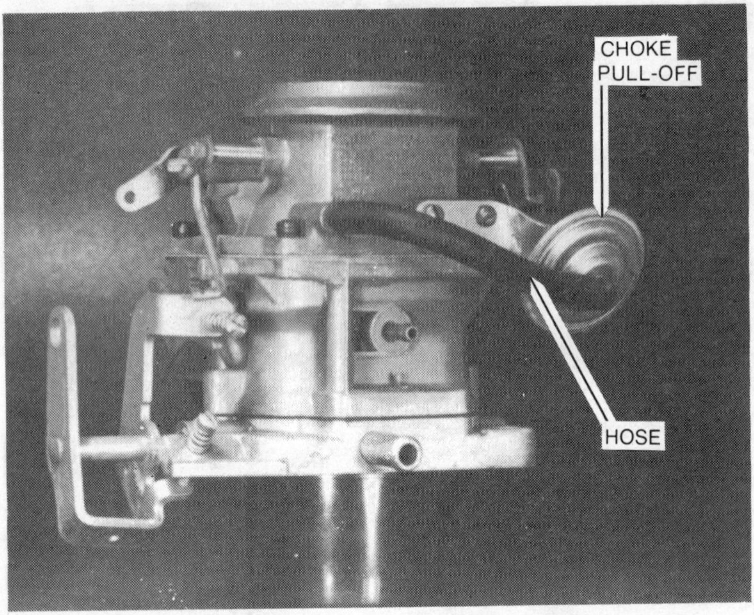

Fig. 7-2. Choke pull-off and hose (optional).

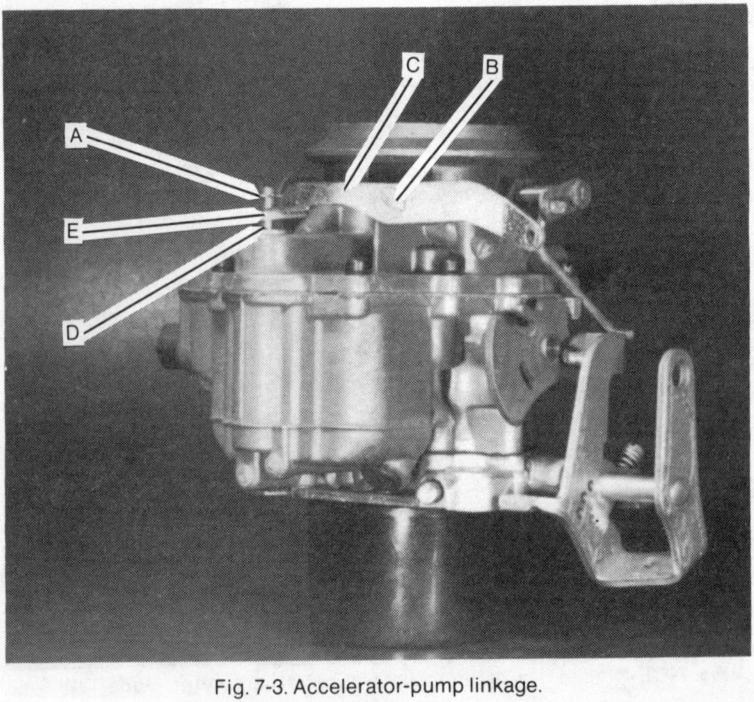

Fig. 7-3. Accelerator-pump linkage.

When carburetor is removed from engine, follow tuneup steps listed below.

To Disassemble

Step 1. Remove retainer clip A (7-3).

Step 2. Remove step-shaped screw B (7-3). *Turn this screw clockwise to remove*. It has left-hand threads.

Step 3. Step 2 released accelerator pump lever C (7-3). Remove this lever.

Step 4. Remove small flat washer D (7-3) from accelerator pump shaft E.

Step 5. Remove clip F (7-4) from linkage G; disconnect linkage.

Some models of the WW have a choke pull-off as shown in Fig. 7-2. Remove pull-off and hose—replace hose with a new one (5 in. of $^{3}/_{16}$ in. vacuum hose).

Step 6. Remove the six bowl-cover screws H (7-4); lift cover from carburetor; discard gasket.

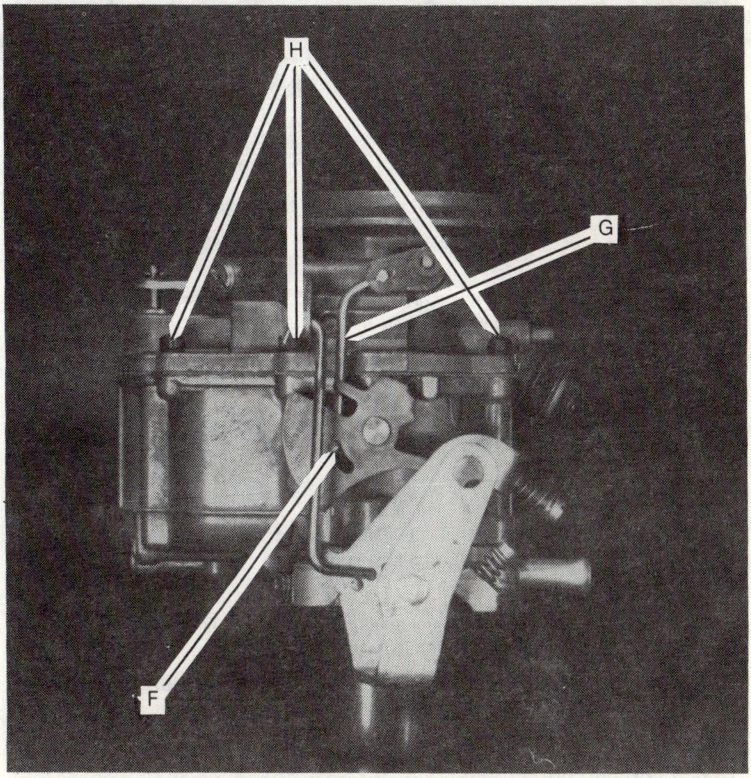

Fig. 7-4. Fast-idle linkage and cover screws.

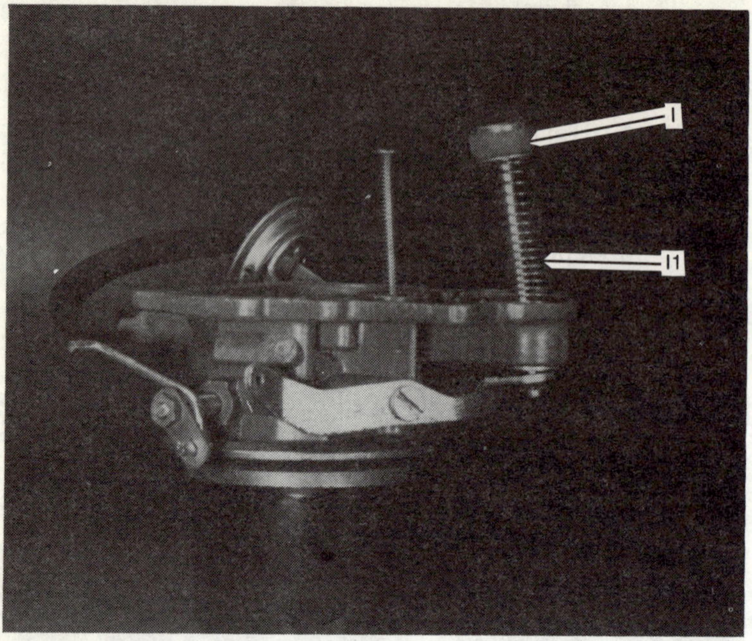

Fig. 7-5. Bowl cover.

Step 7. Remove accelerator pump plunger I and spring I1 (7-5).

Step 8. Remove steel check ball I3 from well I2 (7-6).

Step 9. Use $^9/_{16}$ in. wrench; remove needle valve J and seat K (7-6).

Step 10. Remove spring retainer L2 (7-6) from float-hinge pin L1. Remove pin and float L.

Step 11. Remove main metering jet M (7-6).

Step 12. Lift nozzles N (7-6) out of wells N1.

Step 13. Remove screw O1 (7-6).

Step 14. Remove discharge-nozzle assembly O (7-6). In well P, beneath the screw O1, is a tiny steel check ball P1—remove it.

Step 15. Remove, from underneath, the four screws Q (7-7) that hold the base to the bowl. Separate base from bowl; discard gasket.

Step 16. Remove the two idle-mixture screws R (7-7) and springs R1 from carburetor base.

Step 17. Remove the two large screw-type plugs S (7-7) from bottom of carburetor bowl.

Note: Place all metal parts in bath. Allow all parts placed in bath to be properly cleaned in accordance with directions on container before beginning reassembly. Then remove all parts from

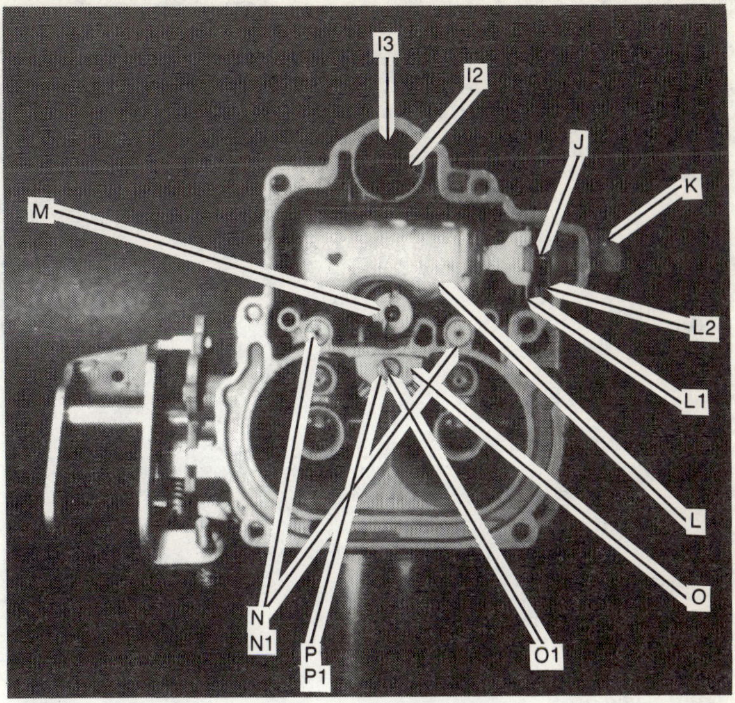

Fig. 7-6. Bowl assembly.

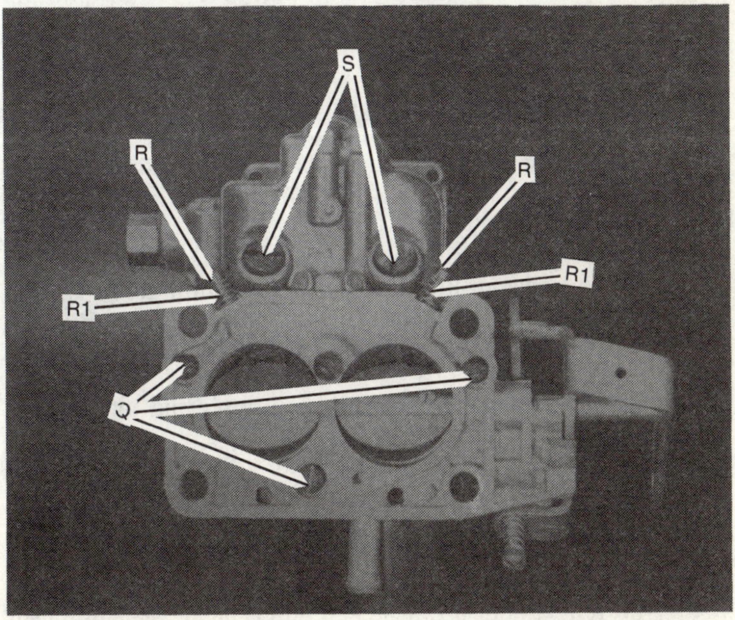

Fig. 7-7. Base-to-bowl screws, idle-mixture screws, and plugs.

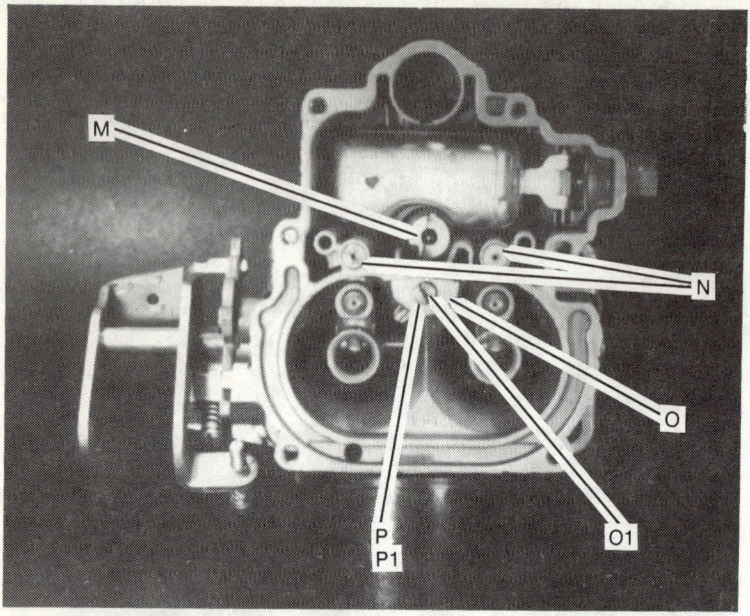

Fig. 7-8. Bowl assembly, showing pump-discharge parts.

cleaner, rinse with water, and allow to dry naturally. Do not use compressed air.

To Reassemble

Step 18. Replace the two idle-mixture screws R and springs R1 (7-7).

Step 19. Use new gaskets; replace the two screw-type plugs S (7-7).

Step 20. Use new gasket; assemble base to bowl. Tighten screws firmly.

Step 21. Drop idle nozzles N into wells N1 (7-8).

Step 22. Drop the smallest check ball P1 into well P (7-8).

Step 23. Use new gaskets under head of hollow screw O1 and under discharge-nozzle assembly O. Replace discharge-nozzle assembly into well P (7-8). Tighten screw firmly.

Step 24. Use new gasket; install main metering jet M (7-8). Tighten firmly.

Step 25. Use spring retainer L2 (7-9) to install float L with hinge pin L1.

Step 26. Install new needle valve J and seat K (7-9).

Step 27. Adjust float level according to instructions enclosed in parts kit. *Float level adjustment is very important.*

Step 28. Drop the large check ball into accelerator pump well I2 (7-9).

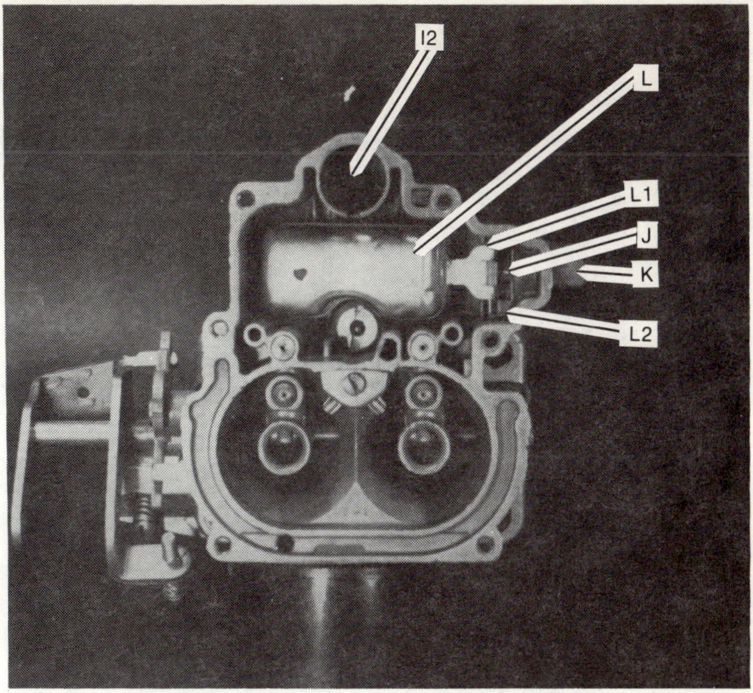

Fig. 7-9. Bowl assembly showing float mechanism and pump well.

Step 29. Place new accelerator pump plunger in well I2 (7-9). Place the long coil spring over shaft of pump plunger—small end down.

Step 30. Use new gasket; place cover on bowl; tighten the six screws H (7-10) evenly and firmly.

Step 31. Place flat washer D (7-10) over accelerator plunger shaft E.

Step 32. Hook lever C (7-10) into linkage G and attach lever with screw B. *This screw has left-hand threads*; turn counterclockwise to tighten.

Step 33. Replace retainer clip F (7-11).

Step 34. Hook up linkage G (7-11). Use small flat washer D under retainer clip A.

Step 35. Install carburetor on engine. Be sure all lines and linkages are properly connected.

Step 36. Start engine. Allow engine to warm up until choke valve is fully opened.

Step 37. Adjust idle with idle-speed screw T (7-12).

Step 38. Adjust idle mixture with screws R (7-12). Turn each screw clockwise until engine begins to run rough, then back it out until engine runs smoothly. It may be

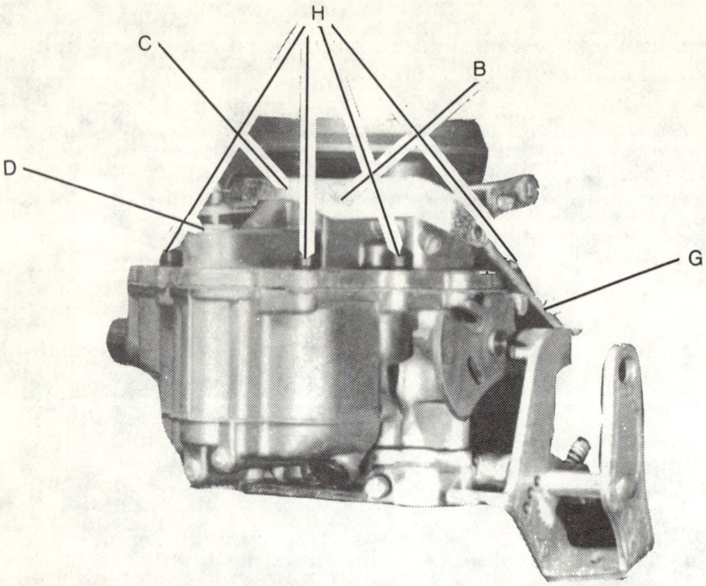

Fig. 7-10. Cover screws, accelerator-pump linkage, and arm.

necessary to readjust both screws, alternately, to obtain smooth engine operation.

Mount air cleaner on carburetor. Be sure element is clean.

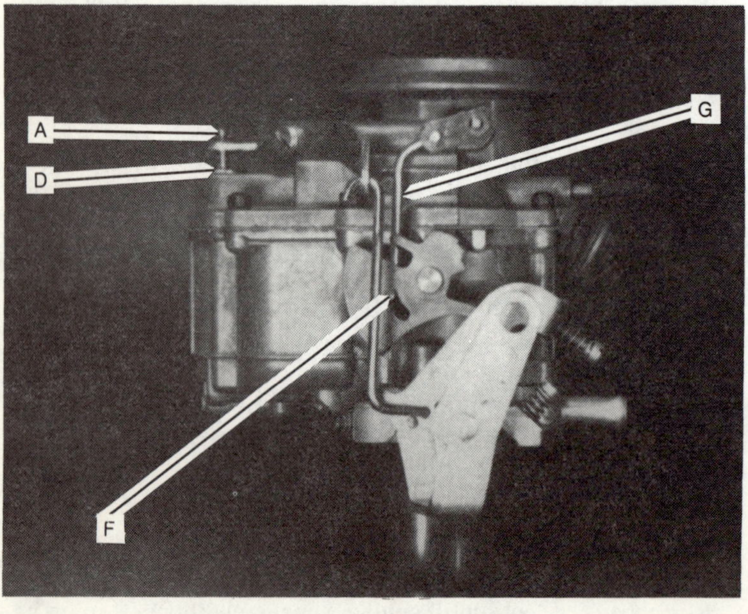

Fig. 7-11. Linkage placement.

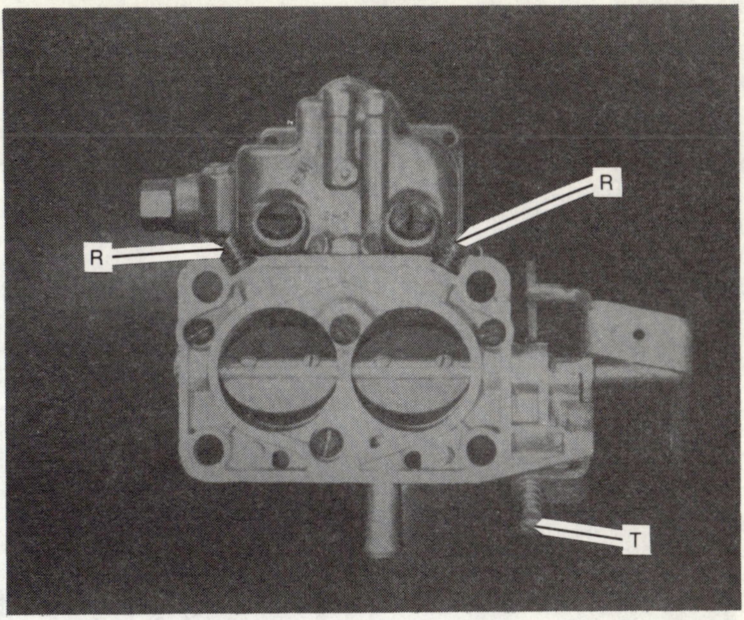

Fig. 7-12. Idle-speed and mixture-control screws.

STROMBERG TWO-BARREL (MODEL WWC)

Please refer to Chapter 1 for a brief description and discussion of carburetor cleaners and the container needed to make up the bath to be used for this task. Tools are listed in Table 7-1.

Preliminary Instructions

Remove air cleaner.

Remove all dirt and grime from exterior of carburetor while carburetor is still mounted on engine; use kerosene and a paintbrush if necessary, but **do not use gasoline**.

Remove carburetor from engine.

For information on carburetor accessories, refer to Chapter 2. If you are not sure you can remember where and how the lines and linkages connect, make notes and sketches in the margin of this book as you remove the carburetor. Figure 7-13, an exploded view, is included to help you identify the various parts.

When carburetor is removed from engine, follow the tuneup steps listed below.

To Disassemble

Step 1. Remove retainer clip A (7-14) and linkage A1.

Step 2. Remove retainer clip B (7-14) and choke linkage B1.

Index
No.

1. Cotter pin
2. Pump rod
3. Cotter pin
4. Unloader link
5. Screw and lockwasher assembly
6. Screw and lockwasher assembly
7. Air horn assembly
8. Throttle body assembly
9. Main body gasket
10. Pump plunger assembly
11. Pump intake check ball
12. Pump plunger spring
13. Air horn gasket
14. Vacuum power piston assembly
15. Screw and lockwasher assembly
16. Bowl vent valve assembly
17. Needle and seat assembly
18. Needle seat gasket
19. Fuel baffle
20. Float pin spring
21. Float pin
22. Float assembly
23. Screw and lockwasher assembly
24. Venturi cluster assembly
25. Venturi cluster gasket
26. Pump outlet check ball
27. Power by-pass jet
28. Power by-pass jet gasket
29. Main metering jet
30. Idle needle valve
31. Idle needle valve spring
32. Main body assembly

33-51. Used on governor-equipped carburetors

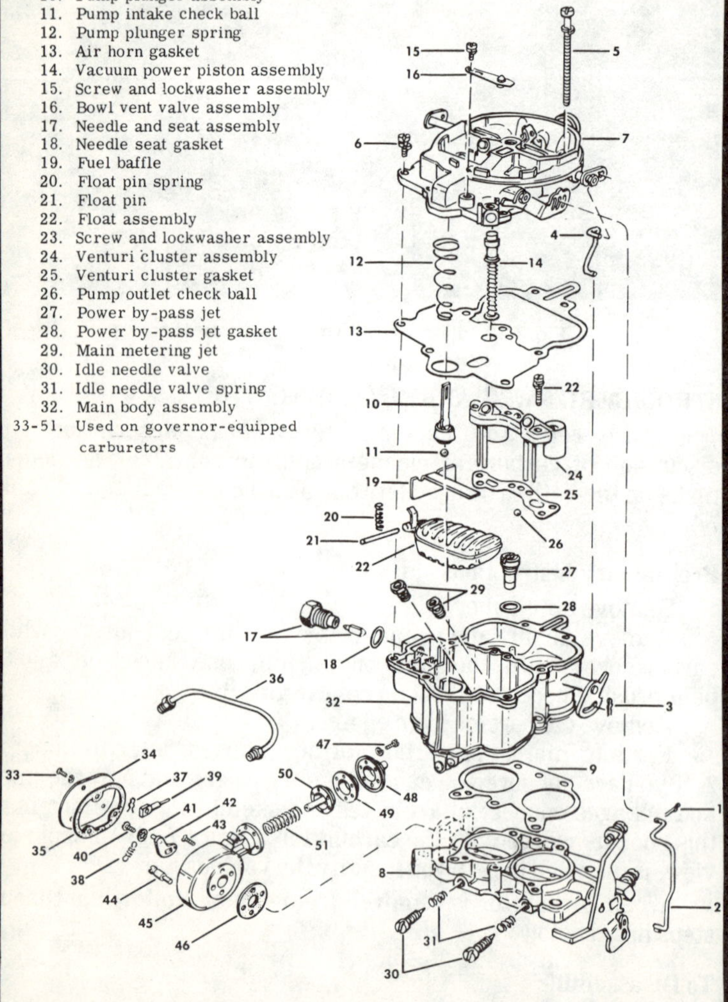

Fig. 7-13. Exploded view of Stromberg two-barrel carburetor, Model WWC.

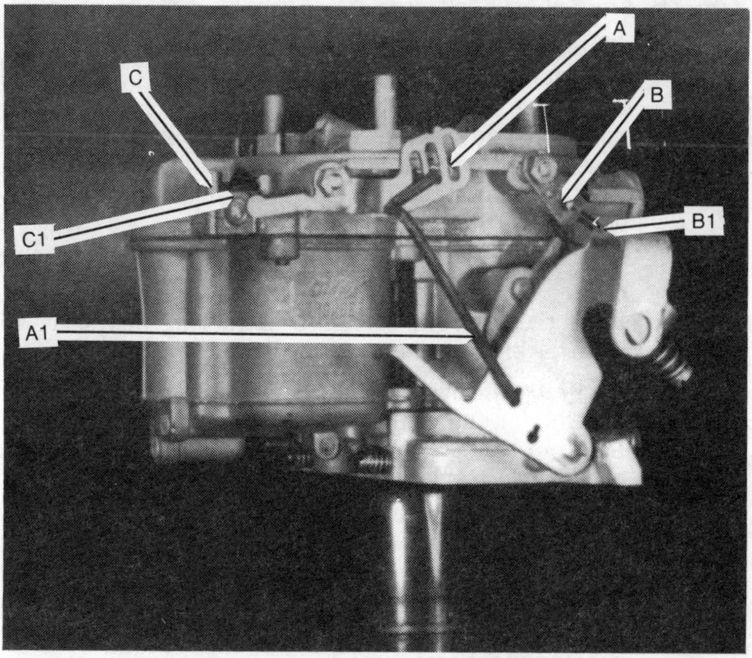

Fig. 7-14. External linkage and bowl vent.

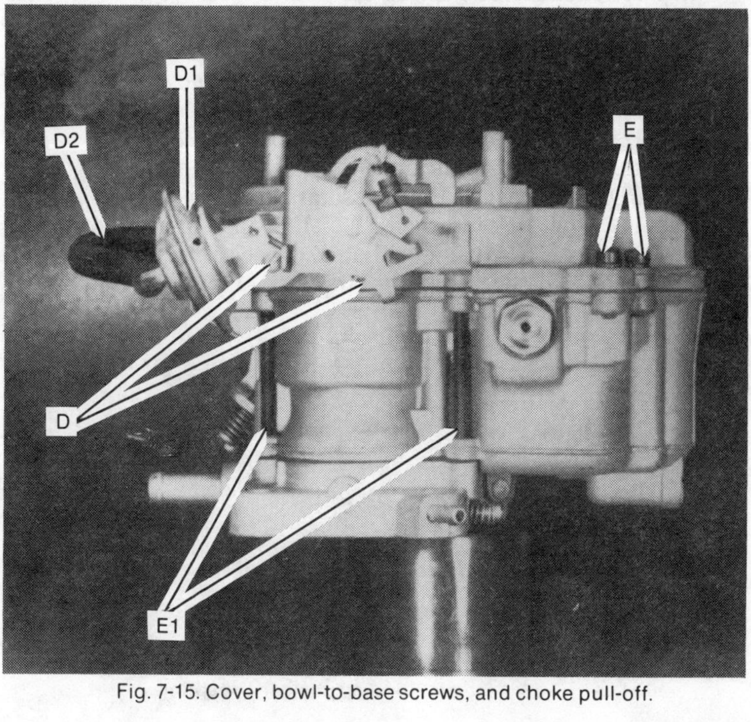

Fig. 7-15. Cover, bowl-to-base screws, and choke pull-off.

Step 3. Remove screw C (7-14), which secures vent assembly C1 to bowl cover; remove vent assembly, but do *not* place it in bath.

Step 4. Remove two screws D (7-15), which secure choke pull-off D1 to bowl cover; remove choke pull-off and hose D2. Discard hose—this hose is not included in your parts kit, so ask for 5 in. of $^3/_{16}$ in. vacuum hose. Do *not* place choke pull-off in bath.

Step 5. Remove seven screws E and E1 (7-15), which secure bowl cover to the bowl.

Note: Four of these screws—E1—are extra long and serve to secure bowl to the base.

Step 6. Lift bowl cover off bowl. Disconnect and remove accelerator pump plunger assembly F (7-16) and spring F1 from well F2 inside bowl cover. Discard plunger assembly. Place spring in bath.

Step 7. Place bowl cover in bath.

Step 8. Remove fuel baffle G (7-17) from carburetor bowl. Place baffle in bath.

Step 9. Remove needle valve I (7-17) and seat I1. Use adjustable wrench to remove seat. Discard both the needle and the seat.

Step 10. Remove coil spring H1 (7-17), then lift float H (with pin H2) out of bowl. Place spring in bath, but do *not* place float in bath.

Step 11. Use broad-blade screwdriver to remove power valve J (7-17) from well J1 in bottom of bowl.

Step 12. Remove three screws K (7-17).

Step 13. Lift out venturi cluster K1 (7-17) and old gasket. Discard gasket. Place assembly in bath.

Step 14. Now, look into well K2 (7-17)—in center of bowl. You will see a small steel ball K3. Look into well F2 where you will see a slightly larger steel ball F3. Place your

Fig. 7-16. Accelerator-pump plunger and spring.

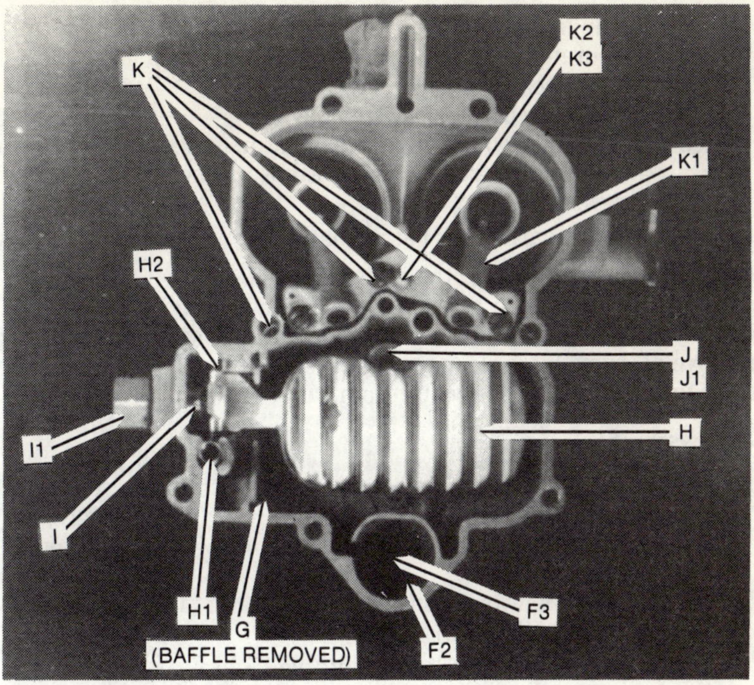

Fig. 7-17. Float, needle and seat, venturi cluster, jet, accelerator-pump check valves, and related parts.

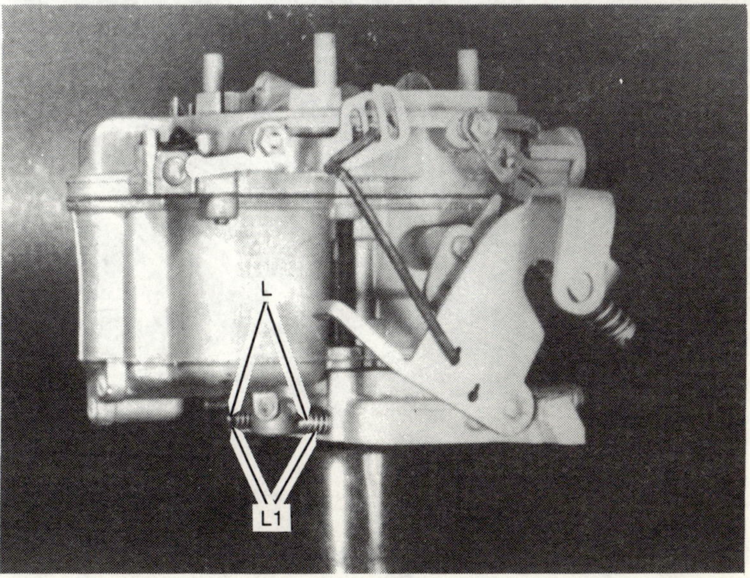

Fig. 7-18. Idle-mixture adjustment screws.

hand in position to catch both balls, then turn carburetor upside down—they should drop into your hand. Discard both of them.

Step 15. Place carburetor bowl in bath.

Step 16. Remove two idle-mixture adjustment screws L (7-18) and spring L1 from carburetor base. Place springs and screws in bath.

Step 17. Place carburetor base in bath.

Note: You have completed the carburetor disassembly. Allow all parts placed in bath to be properly cleaned in accordance with directions on container before beginning reassembly. Then remove all parts from cleaner, rinse them with water, and allow them to dry naturally. Do not use compressed air.

To Reassemble

Step 18. Install the two idle-mixture adjustment screws L (7-18) and springs L1 in carburetor base. Tighten screws until they seat—finger-tight only—then back out 1½ turns.

Step 19. Insert new ball K3 (7-19) into well K2. (This ball is the smaller of the two.)

Step 20. Insert new ball F3 (719) into well F2.

Step 21. Use new gasket. Install venturi cluster K1 (7-19) with three screws K. Tighten screws firmly.

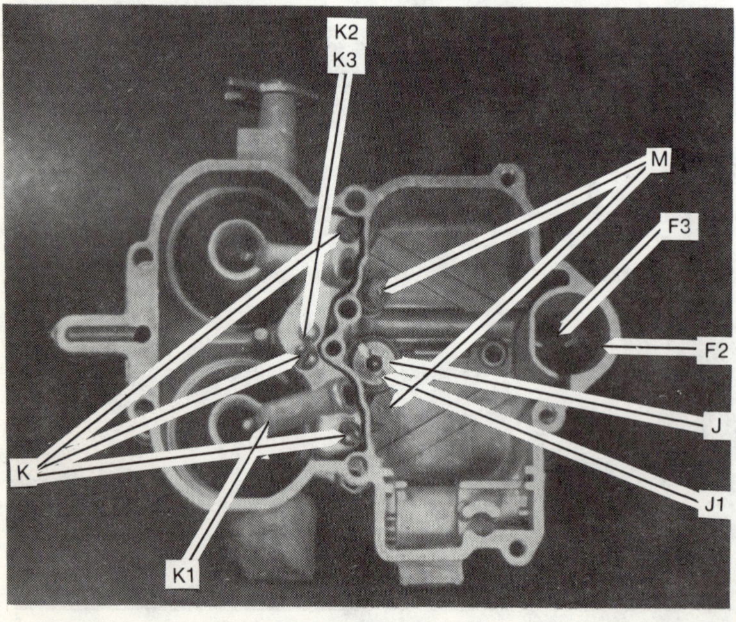

Fig. 7-19. Venturi clusters, jets, accelerator-pump well, and related parts.

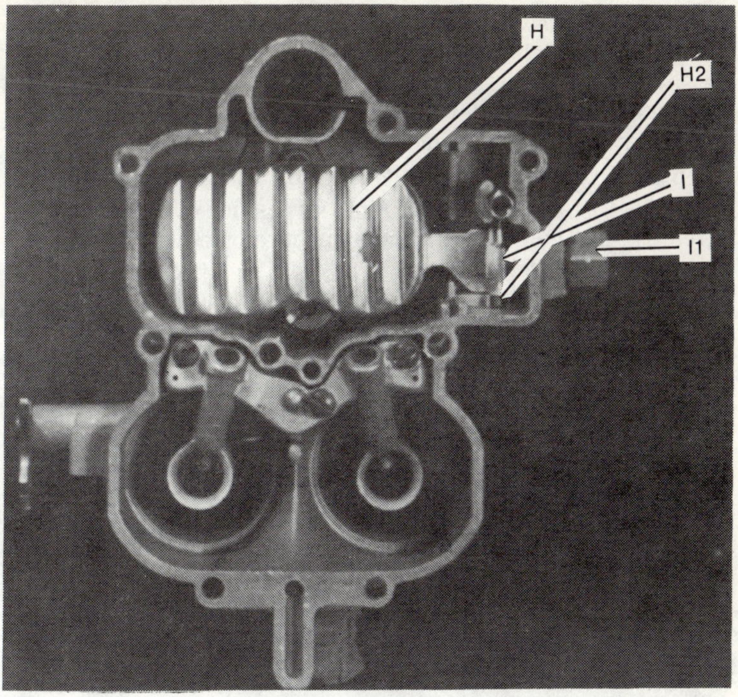

Fig. 7-20. Float assembly.

Step 22. Use new gasket. Install power valve J (7-19) into well J1. Use broad-blade screwdriver to tighten screws firmly.

Step 23. Tighten the two metering jets M (7-19) in bottom of carburetor bowl.

Step 24. Lower float H (with pin H2) into position as shown in 7-20.

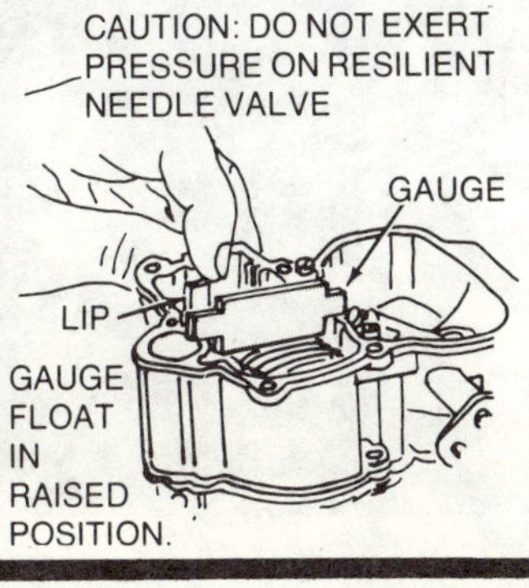

Fig. 7-21. Float adjustment.

Step 25. Use new gasket. Install new needle-valve seat I1 (7-20). Tighten seat firmly. Now, drop new needle valve into seat well.

Step 26. Adjust float level. Refer to instruction sheet in parts kit, and to 7-21. *Be accurate*.

Step 27. Drop coil spring H1 (7-22) into position over float.

Step 28. Install baffle G (7-22) into its position over the float.

Step 29. Use new gasket. Place carburetor bowl into proper position on the base.

Step 30. Place spring F1 (7-23) over new accelerator-pump plunger assembly F, small end first. Compress spring and hook notch in plunger shaft F4 into lever.

Step 31. Place new gasket on carburetor bowl.

Step 32. Lower bowl cover into position bowl. Be sure that accelerator pump plunger shaft F3 (7-23) and vacuum-piston shaft are properly aligned.

Step 33. Insert the four long screws E1 (7-24)—remember, these screws go all the way through bowl to secure bowl to base. Run them in, but don't tighten yet.

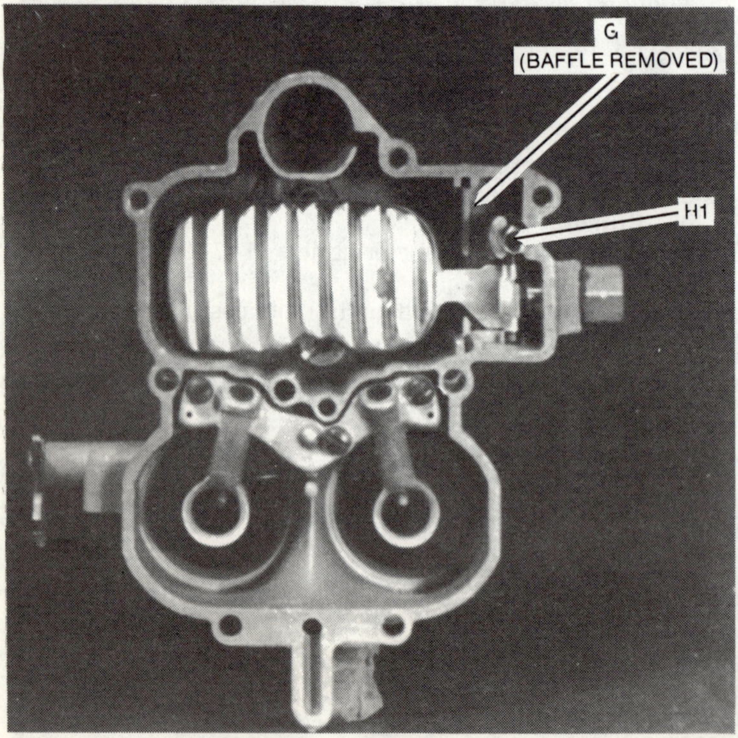

Fig. 7-22. Baffle placement and float spring.

Fig. 7-23. Float cover.

Step 34. Insert the three shorter screws E (7-24). Now tighten these three screws and the four longer ones at the same time, and firmly.

Step 35. Place carburetor bowl vent assembly C1 (7-24) into position on bowl cover and secure it with screws C. Tighten these screws firmly.

Step 36. Install the choke pull-off D1 (7-25) with two screws D. Leave screws loose for the moment. Install new hose D2.

Step 37. Install choke linkage B1 (7-26) with retainer clip B.

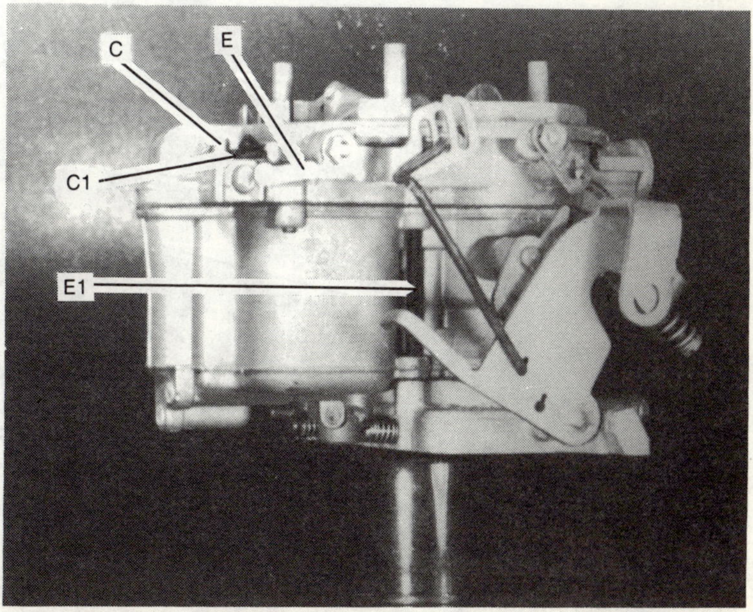

Fig. 7-24. Bowl-cover screws, cover-through-base screws, and bowl-vent assembly.

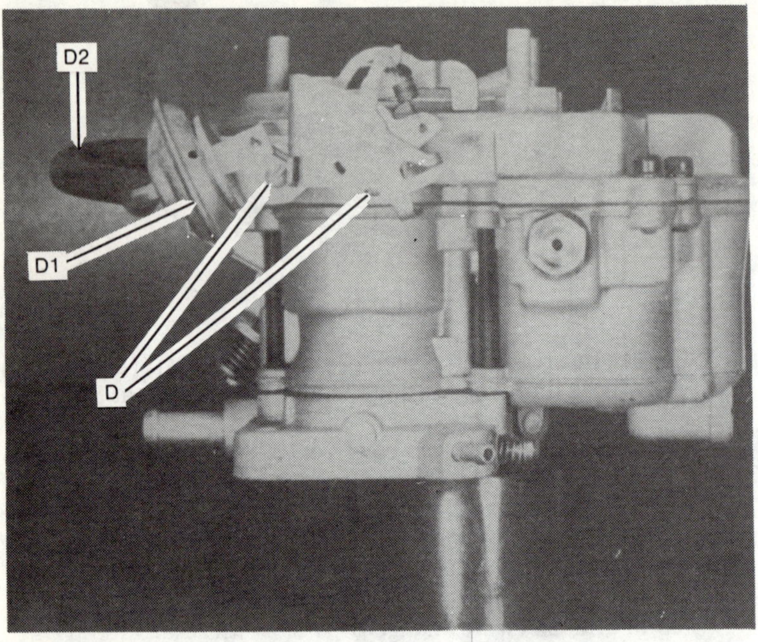

Fig. 7-25. Choke pull-off assembly.

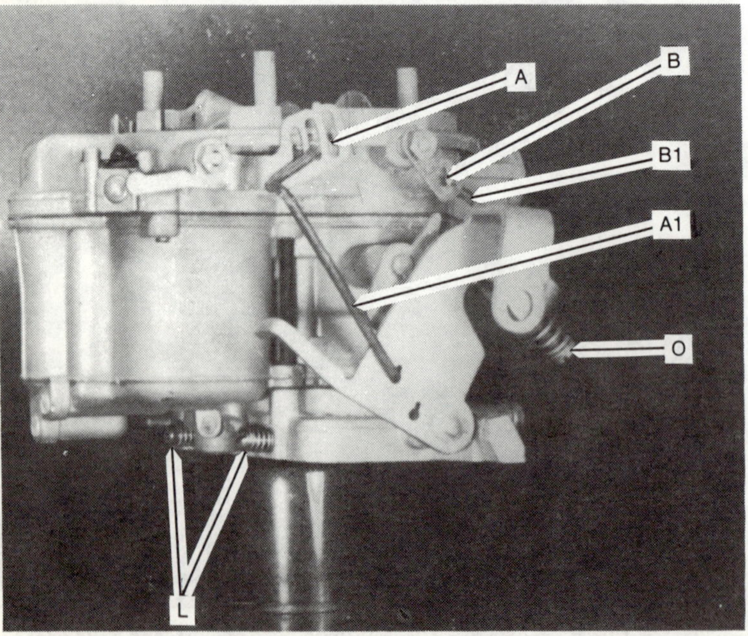

Fig. 7-26. Linkage and adjustment screws.

Step 38. Now, tighten two choke pull-off screws D (7-25).

Step 39. Install linkage A1 (7-26) with retainer clip A.

Step 40. Mount carburetor on engine. Be sure fuel line and all linkage are properly attached.

Step 41. Start engine. Allow it to warm up until butterfly (choke) valve in carburetor throat is in fully open position.

Step 42. Adjust engine idle to desired idle speed with screw O (7-26).

Step 43. Adjust idle mixture with screws L (7-26). Turn each screw clockwise until engine begins to falter, then counterclockwise until engine idles smoothly. Adjust screws alternately until idle is smoothest.

Replace air cleaner on carburetor. Be sure element is clean.

Index

Index

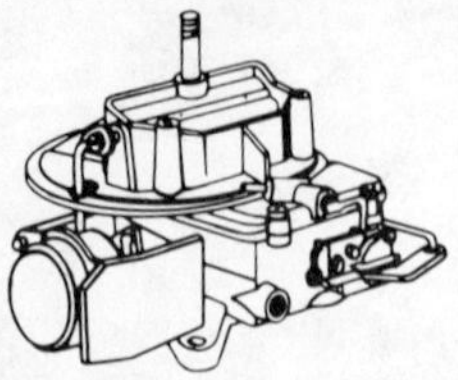